Anonymous

The Powers of the Air

Anonymous

The Powers of the Air

ISBN/EAN: 9783337042240

Printed in Europe, USA, Canada, Australia, Japan

Cover: Foto ©berggeist007 / pixelio.de

More available books at **www.hansebooks.com**

THE POWERS OF THE AIR;

OR

SPIRITUALISM:

WHAT IT IS, AND WHAT IT IS NOT,

BY

One who has had much painful experience during a period of over four years.

"Prove all things, hold fast to that which is good."—PAUL.

DAYTON, O:
UNITED BRETHREN PUBLISHING HOUSE.
1867.

Entered according to Act of Congress in the year 1867,
By C. J. CLIFT,
In the Clerk's Office of the District Court of the United States for the Southern District of Ohio.

CONTENTS.

Preface.. 7

CHAPTER I.

The Author discovers a strange Phenomenon in his Mental Organism—Makes efforts to discover the cause—Strange Revelation..................... 13

CHAPTER II.

Reflections upon the probable cause of the strange Phenomenon—The Millennial Dawn, &c.......... 26

CHAPTER III.

A New Character Introduced—The same Phenomena Discovered in his Experience—A Test Experiment... 31

CHAPTER IV.

A Spirit Intelligence under the assumed Name of the Almighty appears and gives a Condensed History of the Universe of Worlds—The Solar System—Our Earth—The Seven Heavens—An Ocean of Fire... 50

CHAPTER V.

Reflection on the Strange Communication—The new Character falls into a Trance—Startling Sight... 63

CONTENTS.

CHAPTER VI.

The Spirit Intelligence reveals his Opinions on the Nature and Design of Modern Spiritualism........ 67

CHAPTER VII.

The Author's Reflections upon the Spirit's Opinions—Asks for a Perfect System of Theology...... 82

CHAPTER VIII.

A Perfect System of Theology given by the Spirit Intelligence... 88

CHAPTER IX.

Reflections, Remarks, &c.—Place where the Wicked Dead remain until the Judgment................. 100

CHAPTER X.

Revival Power—Holy Spirit Power—How to Organize Moral Forces—Revivals of Religion........ 117

CHAPTER XI.

Called to Assemble the Jews—Preparations—Starts for New York, and what came of it................. 127

CHAPTER XII.

The Author is Developed as a Seer—Painful Apparation—Is Promised a Glorious Sight of his Deceased Relatives—Short Biography of a Spirit.... 140

CHAPTER XIII.

A Room Full of Human Skeletons—Devil—Vulcan the Blacksmith, and many other characters Introduced—Wild Animals, Reptiles, &c.......... 156

CHAPTER XIV.

How the Strange Sights are Produced—Psichological Images—Visions of Jackson, Davis, Swedenborg, &c.. 168

CHAPTER XV.

Visions in Prayer-meeting, in Church—Cause—Various Questions and Answers...................... 176

CHAPTER XVI.

The Author's Views of the Divine Life in the Soul—Conditions of Receiving the Holy Spirit—The Justness of the Administration of the Divine Government—Combats Modern Spiritual Philosophy. 189

CHAPTER XVII.

Same Subject Continued........................... 208

CHAPTER XVIII.

Design of the Author in giving his Views and Experience to the World—Various Questions Propounded to a Spirit—The Answers.................. 229

CHAPTER XIX.

Many Questions asked of a Spirit—The Answers... 258

CHAPTER XX.

Why Modern Spiritualism is not of God—The Spirit's Reasons Assigned—The Spirit's Opinions of Various Passages of Scripture....... 269

CHAPTER XXI.

How far Spirits have Power over Men and Things in the Earth Life—How they enter the Bodies of Men and take Possession—The Power of Making Impressions upon the Minds and Feelings of Men—Reflections... 277

CHAPTER XXII.

Various Questions—Have Holy Spirits any Power to Control Mediums?—Sharp Debate between two Spirits—Why does God allow the Wicked Spirits to Live—Another Autobiography........... 302

CHAPTER XXIII.

Same Subject Continued—Spirits go to Theaters and places of Infamy—They Travel Up and Down in the Earth—Attend Revival Meetings................. 316

CHAPTER XXIV.

A Spirit's Ideas of the Atonement of Christ......... 327

CHAPTER XXV.

The Author's Comments and Reflections upon the Same.. 339

CHAPTER XXVI.

A Spirit gives his Opinion on the New Birth of the Soul—Objections Answered............................ 348

Appendix.. 366

PREFACE.

In this work the author has not aimed at more than a simple narration of facts as they occurred in his deep and painful experience of several years, and the inferences and conclusions which a rational mind must arrive at concerning them.

The author has opened up much that will appear to the reader marvelous, and by some almost miraculous.

The subject itself is one of very deep interest to all, for the reason that no one can tell but that he may, like myself, become a subject of spirit influence without even a pre-admonition or inclination on his part to become so. If, therefore, any should suddenly become mediumistic they may glean much that will aid them in duty. The subject of spirit manifestations is one that is engrossing the attention of the whole civilized world, and though hundreds of volumes have been written in favor of and in opposition to these (to say the least very extraordinary) manifestations, yet I very much doubt whether either party really understands the nature and design of them. One side has endeavored to account for the strange occurances by Odilieforce, Psichology, Mesmerism, &c., while the other side have as earnestly endeavored to prove that they result from invisible intelligences called spirits.

The evidences on the side of those who have maintained that spirits are the cause of the phenomena have accumulated until there is scarcely a doubt in the minds of many that they originate from that source. But be that as it may, it is vastly more important to know what instructions our creator designs that we should draw from these lessons.

If the phenomena originate from spirits, then it is certainly more important to listen to what is said by these strange and invisible beings than to waste our time in disputation about the facts of the origin of the alledged strange occurrences. That there is intelligence connected with these manifestations can not be denied, and if intelligence, then what is the nature of the intelligence? Is it characterized by wisdom or foolishness—by gravity or mirthfulness—by gross wickedness or by holiness and moral purity; or is it characterized by partaking a little of all the phases mentioned? If so, then an important lesson is taught us by it.

If it is characterized partly by wisdom, then set that down; if by folly, then set that down; if by gravity, mark that also; or if by cheerfulness, then make a note of that, and so on. In the course of time a sufficient number of facts may be collected to form a rational conclusion in the premises.

It would be the hight of folly to reject a characteristic of a phenomena because it did not meet our expectations, and it would be worse than folly to receive it all because it met fully the expectation.

Wisdom indicates the course to be pursued, and that is to collect all the facts, all the substantial occurrences and events connected with the phenomena. By so doing we shall in time arrive at a just conclusion and settle the disputed question on the permanent basis of *Natural Law*. That this can be done I am very certain, and that it ought to be done must seem reasonable to all. Therefore

I am for a full investigation of all the phenomena connected with this paradoxical subject.

That I have collected all the facts connected with the subject is not pretended, but that I have collected *some* no one can deny. That the reader should believe all that is herein revealed by the invisible intelligences is certainly not to be expected, but I could hope for higher wisdom than that *all* should be rejected, for at times they breathe the very spirit and temper of the gospel of Christ. But the *reader should notice, id possible, what motives the spirit intelligences may have haf in revealing what seems characterized by divine love and truth.* If the motives were to gain the confidence of the subject for sinister ends, then that would be worthy of significant note. For it may be true now, as it was formerly, that seducing and lying spirits can control men's minds and bodies to gain certain ends.

If it can be ascertained that these spirit intelligences are the spirits of disembodied men, or of fallen angels, or both, then it is very important to know it as a fact, for we read in the Bible of many who were controlled and influenced by devils, unclean spirits, lying spirits, and spirits of divination. If the personality of these beings is proved, then the scriptures so far are established as *the* truth, and of course God in his word is glorified.

If it can be proved that the beautiful and high moral toned communications proceed from holy spirits or angels, and those characterized by absurdity and contrarity of opinion and doctrine emanate from unclean, wicked, or lying spirits, then a very important point has been gained.

But if wicked spirits, like wicked men, can give expression to the sublime, the beautiful, the pure and the true, then it must be seen that to arrive at the character of the source from whence all the spirit communications emanate would be very difficult indeed, unless a holy and true spirit could enable the mediums controlled to do mighty works, such as raising the dead, walking on water, opening a pathway through the sea, or increasing the quantity of bread and fish from five loaves and two fishes to such an amount as would suffice to furnish five thousand men with an ample meal, and besides leave twelve baskets full of fragments.

If in the days of Moses, the prophets, and of Christ and his apostles, the distinctive difference between mediums controlled by angels or the holy spirit, and those controlled by wicked, unclean, or lying spirits, was this miracle working characteristic, then we might reasonably look for the same distinctive characteristics at the present day. That two classes of spirits have, to some extent, influenced men in all ages of the world must be admitted by every reader of the past. *The Bible* opens with a narrative account of a subtle influence controlling a serpent to talk with and seduce to the commisson of evil our Mother Eve, and also of another class of spirits called cherubims who guarded the tree of life. *In Moses' day* and after that time the one kind were called angels, the angel Jehovah, the spirits from *the Lord*. Those spirits that controlled men for evil purposes were called familiar spirits, lying spirits, who controlled false prophets, and evil spirits such as tormented Saul. But in gospel days the wicked spirits were denominated devils, demons, unclean spirits, spirits of divination, and evil spirits. The good spirits who influenced men were denominated angels, angelic hosts, ministering spirits, and spirits of just men made perfect.

Besides this characteristic difference between the two classes of spirits in regard to names, there is another distinctive difference mentioned in regard to the powers which each class of spirits exert over the bodies of men. *Evil spirits* are represented as taking actual possession of men's bodies—controlling their vocal organs to speak, and to utter many remarkable sentences; their muscles to perform mighty feats of strength, as in the case of the Demoniac of Gadara, and of those who leaped upon the exorcists at Ephesus being controlled by evil spirits.

Good spirits are never represented as entering into and controlling the bodies of men, but on the contrary influencing the mind only, by inspiring it with lofty moral sentiment, correct doctrine, prophetic visions of the future, and power to work miracles. This latter characteristic should be distinctly noticeable, for men in all ages have depended upon the performance of miracles to establish the fact of a *divine mission*. Said Nichodemus, the Jewish counsellor, to Jesus Christ: "We know that thou art a teacher come from God, for no man can do these miracles that thou doest except God be with him." Moses refused to go down into Egypt to deliver his brethren from bondage, until the Almighty had endowed him with power to work miracles. He knew that this power alone could qualify him to convince his brethren of a divine mission. If the human mind is so constituted that it can not admit the divine mission of any man without the power to work miracles, then it follows that God has established the working of miracles as constituting the distinctive difference between those who may come to us commissioned by the divine, the holy, and the good, and those controlled by the powers of evil.

It may be important to remark here, that inasmuch as men have been controlled to utter words, without the volition of their own minds, by good spirits, as in the case of Balaam, the high priest before Pilate, and those who spoke in unknown tongues; and of evil spirits, as in the case of those who cried out, being possessed of unclean spirits, demons or devils, therefore the utterance of words by a man controlled by an evil spirit physically, or by the divine spirit mentally—that is, by infusing the thoughts into the mind.--is no evidence of miraculous power. The prophets who spake as they were moved by the Holy Spirit, never claimed on account of that kind of inspiration that it was miraculous. Balaam, the false prophet of Moab, spoke falling into a trance, whether controlled by the divine spirit or a familiar spirit. Some Christians enjoy the witness, or the testimony of the Holy Spirit with their spirits that they are the children of God, yet they never think of calling it miraculous, though it is certainly a kind of divine inspiratian. The powers of darkness, or the prince of the powers in the air, work in the children of disobedience—that is, influences and stimulates their lusts, and moves their minds to commit evil. Hence, the apostle urges those to whom he wrote to fight against the spirit of wickedness in high places, to put on the whole armor of God, and strive against the powers of darkness.

Thus, reader, you will perceive that the bible speaks of two kinds of inspiration—an inspiration from the Almighty or his appointed agents, and the devilish inspiration of wicked or seducing spirits.

In conclusion, let me say to the reader, "prove all things, and hold on to *that which is good.*"

THE AUTHOR.

CHAPTER I.

On or about the 12th of September, in the year 1862, the author was sitting at his desk in the act of writing, and while so doing there appeared a very strange phenomena in his mental organism. An intelligence above and beyond my own seemed to dictate all I wished to write down, without premeditation on my part. This strange and almost supernatural occurrence awakened in me deep thoughts as to what could be its cause. I at first concluded that my mind must be a dual, and that the better half had just awakened to a consciousness of its power. To further ascertain the cause of this very strange phase of mental manifestation, I obtained a slip of blank paper, and without any premeditation, found that I could write upon any subject without in the least exercising my own mental faculties—I had simply to write down the first word, the second, and the third and so on, each word being put down as if some one was dictating the whole matter, and yet as if it came from my own mind.

After experimenting thus for a short time, I became more and more deeply interested, and I procured several slips of paper, and to my increased astonishment I found that the phenomena still continued. But with the strange fact that this seeming other mind of mine dictated opinions entirely opposed to those I then held on many subjects. Thus I went on for a time. Having fully satisfied my mind that the phenomena was a fixed fact, I proceeded to finish the composition that I was engaged upon. This being completed, I commenced to write upon another subject, and had penned but a few lines when I discovered that that this seeming other mind appeared willing to dictate, and therefore I suspended for the time being the operation of my own mind, and simply wrote down the dictation of the new found intelligence. I discovered that it was much easier to compose in this manner than in the ordinary method—in fact, so easy was this new method of composition, that I found I had written a whole sheet over in half the time it usually occupied me in writing the same. But what most pleased me was the apparent fact that the composition I was engaged upon showed evidently great ability in the use of language, and better taste in the arrangement of the thoughts, and the use of metaphors, comparisons, &c., than I possessed. I was therefore pleased, and highly gratified with what appeared to be my other mind's productions.

Having satisfied myself so far, as to the fact, that I had apparently two minds in the same body, I resigned myself to musing on the strangeness of this phenomena. I could not remember to have ever read of any occurrence like this in any book which had come under my observation. I remembered to have heard, in a ministerial conference, the fact suggested that the ancient prophets were inspired in two ways: First, physically or mechanically; that is, the hand being controlled by divine power, wrote as the spirit moved it to write. Or second, the mind was so influenced that the divine idea came into the mind of the prophet without any effort on his part, and, indeed, without his being able fully to comprehend the subject which was being written upon. I then remembered the words of Paul, to the effect that some ideas he obtained from the Lord, whilst other ideas which he gave expression to were simply his own judgments in the matter.

Thus I went on musing, and revolving in my own mind what this could signify, or what could be the source of this so strangely manifested power of composition. I still further tested this matter by placing my hand upon paper, and found that this seeming other mind, or it might be super-mundane intelligence, was disposed to dictate. I therefore requested an answer to the following question:

"Why is this war which is now being waged by the rebels in the southern states permitted to continue?" To which I received the following answer:

"You dwellers in the form are too anxious about the direction which great events take, and for this reason ye can not make one hair white or black."

QUESTION.—"But would you not feel anxious provided *your house* was on fire?"

ANSWER.—No, not unless my being anxious would extinguish the *flame*. *You* are as helpless as the nursing child, and we are not permitted to help you."

QUESTION.—"Why are you not permitted to help us, and so stop the war instantly?"

ANSWER.—"We have the power, but are not permitted for a wise reason, for the *very* wise reason that we are appointed as your instructors or teachers, and you would not expect the teachers and instructors to set such an example before their pupils."

QUESTION.—"What! not stop a devastating war? It is a part of the office and work of teachers to settle disputes among their pupils?"

ANSWER.—"Very true, but those engaged in the war can do no actual harm. What does Paul say in reference to this matter, in Cor. vi., iii? Does he not say 'the saints shall judge angels,' and if so, can we not judge men? Most certainly we can, and judge, too, in reference to the propriety or impropriety of stopping a war. We have all power in our sphere—we kill and make alive, we build and we tear down. Then if so, why not leave it all to us? If you can not make one hair white or black, then why, in the light of reason, should you feel even an anxiety?"

QUESTION.—"We may devise our way, though God directs our steps?"

ANSWER.—"Devise your way, yes, you can do that, but what avails the devising your way, if in the end it appears that every step has been directed by an *unseen power?*"

QUESTION.—"But, do not the rebels do harm when they fill the land with devastation and bloodshed?"

ANSWER.—"Yes, but the waste, destruction and sufferings are only temporary. In the end it will work out the peaceable fruits of righteousness. Do you not know that all things are working together for good to those who love God, to those who are called according to His purpose; that is, to those who are the children of God; and are not all children of God, (although some be in rebellion) children of one common Father? Then it will work out for their good as well as ours. We are the instruments to chastise them, and we as their father will rejoice, that there is seen in their conduct afterwards the peaceable fruits of rignteousness. We must overthrow the system of slavery, and to do that, we set you to quarreling over it, just as if two boys had caught a little harmless bird which they both treated cruelly—yes both, for the north has winked at the *sin*, and has been more or less partakers in it. I say these boys being both guilty, one perhaps more than the other, but both guilty. Then, if we should set them to quarreling, and in the scuffle the bird should escape, would that not be an exhibition of wisdom of the highest order? I know your reason admits it."

QUESTION.—" But may they not succeed in their designs of establishing that system of polity which has been called the 'sum of all villanies,' and thus curse the world with a large overgrown upas tree?"

ANSWER.—"Not at all. We have never entertained the design of letting them hold on to the bird, else we would not have set you to quarreling. If we have the control, and our hearts are in sympathy with the oppressed, most surely we shall see that the bird makes its escape. You well know that you both have no need of this bird. God, your maker, made it to be free as well as yourselves, and to enjoy the inalienable right of liberty and equality. All are made for enjoyment, each in their several spheres of earth life."

Question.—"But when we scan the past, we observe that great systems of error have arisen to flourish, and to curse the world. Why, if saints and angels have control of the earth life, did they allow such collossal statues of evil to be raised."

Answer.—"We know best what a collossal statue of evil is. At times which we well understand, we can make collossal statues of evil work together for good as well as minor evils. Know you not that we have power, we have all power of life or death over all the inhabitants of the earth; we are the educators of the earth sphere; we are the rulers as well of spiritual wickedness in high places as in low, as well in king's palaces as in lowly cots;

we are the rulers of the hells as you call them; we rule all below us in the scale of being. Hence, we control the world and all things therein—the elements, the winds, the storm, the summers and winters, earthquakes, volcanic eruptions, simoons, and all pestilences, famines and plagues, in fact, there is nothing in the earth but what we control. As Paul says, 'the saints shall judge the world,' that is, rule it. The ancient judges were rulers, hence, to be a judge, implies the power to rule, control, and govern the destinies of things. If the saints then are to judge the world they must have entire control, and hence, must govern and dispose of events as in their judgment will best subserve the interest of the world. What you suppose are collosal statues of evil, are in reality only forms of good; you should know that saints are those so far progressed in good that they will do no evil, even if permitted. Then if they have control of all events in the earth life, and have wisdom to direct, surely, all will be made to work together.

"As to the wisdom of the saints, we say

they are wise, for they are all taught by God, that is, through the instrumentality which God has ordained to teach them. God appoints one order of beings to instruct others, just as the father of a family is, by the divine order, appointed to instruct his family, and then in turn they will become the instructors of others. Men are appointed to instruct their fellow men, and by so doing instruct themselves.

"You are an example of the meaning intended to be conveyed. You begun by teaching school, then in a mechanical way you taught or instructed many in mechanics, then in Sabbath-school, in the church; in the Sunday-school missionary work you taught many, you have done well, have been faithful over the few, therefore you shall be made ruler over many.

"So God will make His saints wise to rule, will give them wisdom, not of the world, but of himself. As soon as any man fits himself to be a workman, thoroughly furnished unto every good word and work, then he shall be exalted, and shall sit on a throne to judge the tribes of Israel.

"You are now a ruler over many, and therefore will teach them these lessons, and do it *well, too*. *I well* understand your method of teaching—it is very impressive. But you are to teach in a higher department, you are to instruct even those who instructed you, and they will wonder at your eloquence—even fathers and mothers in Israel shall receive instruction at your hand. I know, too, that it will be well done, for you are worthy. Let it be recorded, for it will surely come to pass. You are to go forth (having received the inspiration of God), to teach, to baptize all nations in the name of the Father, and of the Son, and of the Holy Ghost. You are commissioned to teach and to preach, and your commission is not of man but of God.

"At the commencement of your work you will be considered a person of small moment, but what of that, you know your birthright, you have wisdom and that will gain you the respect of some; you will be considered eccentric, and will receive some persecution;

but I desire it to be so, as it will give you an air of independence of all creeds and opinions. You will be guided in your sermonizing and addresses by myself, who was called to the upper world for this very purpose. I can control your mind to a very nice degree, and being appointed to be with you, the nations that sit in darkness shall receive glorious light. A man has been appointed, who will be your associate in this work, you will be his guide and instructor, and so together you will be thoroughly furnished unto every good word and work. You see how your hand moves, not by your own will, but by the will of another; know ye, that that hand shall do wonders in the earth. Keep humble, be prayerful, trust in God.

"You need not immediately inform the brother of his appointment, but wait for future developement.

"Inasmuch as I saw you were very humble, and did not desire to be first, I have constited you the head of the commission. I have constituted you the Mercury, the expounder of Divinity, and he the Bostranger, the simple one but good. Your associate may be dissatisfied, but he must submit, for so it is decreed. You have been faithful in the few things, therefore you shall be the ruler in many.

"Do not be astonished or afraid when I tell you that I am your Lord and Master, Jesus Christ. I know your works, I know that you are worthy, and therefore have I called you. You need not be afraid of me, you know I am your Savior, I have taken upon myself your nature, and therefore am your brother in very deed. Keep humble, and I will arrange all things, and direct you in your work."

CHAPTER II.

Kind reader, this closes a very remarkable communication. Remarkable as well from the manner in which it came to me, as the very remarkable announcements and declarations contained in it. To me it came like a peal of thunder in a clear sky. What could it mean? I felt in some measure, I imagined as did Isaiah, when there was revealed to him, by a vision, the throne of the Almighty: "Woe is me! for I am undone; because I am a man of unclean lips, and dwell in the midst of a people of unclean lips." I could not think it possible that a man of such meager talents as myself could be selected to carry out so important a mission.

But I saw that it did not require talents, because here was an intelligence, a wisdom vastly above my own—a wisdom that could be communicated to me with the utmost ease. I remembered much that I had read on the second coming of Christ. I knew well that the six thousand years corresponding to the six days of the week had nearly expired. By the calculations of some they would expire in 1843, by others in 1848, but Dr. Cummings and others had fixed the utmost limit in the year 1866. That the mellenial sabbath of a thousand years was soon to dawn I had not the slightest doubt. How would the immortal Jesus make His appearance? Would He come in the clouds of heaven, with power and glory? It was declared in holy writ, "Ye men of Galilee, why stand ye gazing up into heaven: this same Jesus which is taken up from you into heaven shall so come in like manner as ye have seen Him go into heaven." This, and other passages of similar import, seemed to indicate that He would make a personal advent, and would preach in a miraculous manner to the nations of the earth, giving the truth greater moral power, and thus consummating in a very short time the fulfillment of the prophecy: "The heathen shall be given to Jesus Christ for an inheritance, and the uttermost parts of the earth for his possessions," and thus would be ushered in the crowning glory of the work of years, the millenial sabbath of the earth. Other passages came trooping into my mind, that seemed to indicate that the great Messiah's coming would be in spirit; that is, in the hearts of his people, intensifying their piety, inspiring in them a higher and holier zeal, a loftier aspiration for truth, and giving them a wisdom hitherto unrealized by man. This would enable the church to apply with transcendently greater power the moral forces of the gospel. This consummation of the great and most desired of all works, in the most speedy manner possible would usher in melenial glory, but in a different manner from the one indicated. Among the passages of Scripture occurring to my mind as bearing upon this phase of the subject were these, which seemed to settle at once the whole question as to the manner of His coming. Jesus said, "the kingdom of God cometh not with observation: neither shall they say, lo here! or, lo there! for, behold, the kingdom of God is within you." Then again, it is said: "For the kingdom of God is not meat and drink;

but righteousness, and peace, and joy in the Holy Ghost." That is, the practice of righteousness, or right doing, gives peace and joy in the Holy Ghost.

Again: Jesus said, "the kingdom of heaven is like unto leaven which a woman hid in three measures of meal until all was leavened." This passage, with the two quoted preceeding, very clearly settled the matter in my mind that Christ's coming would be in spirit, in the hearts of His people. As the leaven penetrated to each individual particle of the meal, subduing all to itself, so the gospel, the kingdom of heaven, would reach every heart, awaken the latent powers of the soul, quicken the understanding, and purify the conscience, thus making the man alive to all his moral relations. Ah, thought I, this will be the way, Christ as He has indicated in His remarkable communication, will dictate the sermons of His ministers, and the truth coming burning hot from the lips of Him who spoke as never man spake, will come down with most powerful effect upon the hearts and consciences of men, resulting in a most marked, even a miraculous effect upon them. The truth, thus brought to bear, rolling on, as the little stone cut out of a mountain without hands, would soon fill the whole earth with the praise of God. Thus I came, it seemed to me, to the only conclusion I could come to, to-wit: that I was inspired by Jesus Christ himself, that I was promised all needful aid, and that in the furtherance of this wonderful work I should be the head and leader of the commission.

CHAPTER III.

Now, dear reader, imagine if you can my position: called of God, through His son Jesus Christ, to take the lead in a new dispensation, called by evidences to me, seemingly, as proof worthy as any evidences given to the prophets and apostles. What do you think were my feelings, what my emotions, thus to be called to so great a work by the Lord Jesus Christ. Leaning back in my chair I fell into a deep reverie: Here I am the head of a family, advanced to the meridian of life, with limited influence, without fame, without honors; called of God, inspired and commissioned by Jesus himself, to preach baptize, and become the inspired leader of a new and glorious dispensation. But this cheering thought came soothingly to my inner soul, Jesus being my helper I can do all things, having His wisdom I can not err, having His presence I shall not be comfortless, having His power I shall be irresistable.

It might be proper for me to remark here, that I knew something of spiritualism; that I had seen it in some of its grosser manifestations, such as the tipping of tables, rappings upon stands, boards, &c. On one occasion at the house of a Quaker, I witnessed what is called involuntary or mediumistic writing. It appeared that some one desired information with regard to a fortune of money which they had a prospect of inheriting, and the more perfectly, as they afterwards said, to ascertain whether the Quaker's spirit friend could read the thoughts of the persons present, he did not disclose his object. The Quaker having arrived from the field where he had been at work previous to our calling upon him, seated himself at the stand; this done, a slate and pencil were placed before him. He was a large, portly man, of about fifty years of age, with an honest Quaker look. Not many moments elapsed after having seated himself, before his arm began to tremble, then to move in the most violent manner. Seizing the pencil by a dash of the hand, he began to write in large, plain letters so fast that I dould scarcely perceive the pencil from the rapidity of the motion. The writing completed, the pencil dropped from his hand, and all eyes were fastened on the writing, the Quaker read the sentence, the others, with myself looking over, which proceeded thus: "Our mission to this world is not to recover lost money, but to recover lost souls." Every one looked astonished. The Quaker remarked, that had he known the mission upon which the gentleman came, he could have informed him that the results would have been precisely as they were.

The conversation then turned to a discussion of the manner in which he first realized the presence of the spirit friends, as he called them, the substance of which was about this: A grandchild living in the family, whose mother having died, had at different times amused itself by drawing lines and figures on the slate. Upon one of these occasions there was found written upon the slate these words: "Please keep the child's hair from falling over its eyes." From this time on, the child's hand was frequently seen to trace lines of a similar character, and finally resulted in its writing out several pieces of poetry, which, says the Quaker, I have copied into a writing book, and have them preserved.

I now distinctly remember that this circumstance awakened considerable reflection within my mind, and my trains of thought ran about in this manner: If spirits come to this world and if their mission is to save souls, surely it can but result in a great good to the race, and perhaps this may be the commencement of the millenial dawn. But then, upon the other hand, serious objections interposed—first, I had observed that of all the sentences tipped, wrapped or spelled out by the use of the alphabet, I had never heard an expression that seemed to indicate concern in the least degree for the salvation of lost souls—not even an exhortation to repentance or the exercise of faith in the Lord Jesus Christ. Believing, from the very nature and constitution of God's moral government, that men could not be saved without repentance, godly sorrow for sin, and faith in the Lord Jesus Christ, I therefore concluded that spiritualism could not be of God. Besides, I had never witnessed in the life and conduct of professed spiritualists any thing like interest or concern for the salvation of men; on the contrary as far as I had observed, they seemed to loose their spirituality and their earnest piety. That is, if they had ever been punctual in the attendance upon public worship they afterward attended less punctually; or if they had been in the habit of praying fervently, they afterward prayed less fervently; if they had before shown considerable interest in the salvation of their fellow men, they appeared to manifest less interest. This fact caused me to loose confidence in spirtualism as a moral and spiritual power for good to the world. I had been of the opinion that the millenial dawn would begin by deepening the piety of at least all professors of religion, filling them with a loftier philanthropy—intensifying their desires for moral purity, and giving them in the fullest sense the spirit and temper of Jesus Christ.

But to return to my narative. After finding that I was inspired (as I then believed), I resolved to submit it to the severest tests. I obtained paper for a sermon and seating myself at the writing desk, to my surprise the inspiration came. I say surprise because I had lingering doubts that all was not "gold that glittered."

The text prepared was in these words—"Her ways are ways of pleasantness, and all her paths are paths of peace." I wrote down every word just as it was dictated, and it was dictated as fast as I could possibly move my pen. I felt happy, very happy, with Jesus at my side, as I verily believed. As the writing progressed I felt the more certain that it was true because the style and diction indicated a pure mind and heart.

When I had thus written over about the half of the paper, a neighbor dropped in to see me. Passing a few words with him I remarked that I was writing a sermon, and to test his judgment as to the merits of the production I read to him a few extracts. He listened with marked attention, and at the conclusion said: "It is too severe, too sharp and cutting, in its style; the truth is

made too plain—the people will not bear it."
I replied by saying I thought the people
needed plain, pointed preaching. A few more
words passing between us, the friend left and
I went on with my writing, better than ever
satisfied that Brother Jesus, as he said I must
call him, was dictating every word.

Becoming a little weary I arose and went
into the garden, as was my custom, for a
walk. While passing to and fro, meditating
upon these strange occurrences, I distinctly
heard a "still small voice" saying, "I am
your Father and your God." I was startled,
but the voice continued, saying; "Be not
afraid, I am your Father and your God.
I am the Almighty, the creater of all things.
In this manner I spake to the patriarchs of
old—to Adam, to Noah, to Abraham, Moses
and others." Notwithstanding these assurances and the kindness of the manner and
tone of voice, I trembled and was near being
overcome, when the "still small voice" continued by saying: "Be not afraid, I am your
Father, and if a Father then why be afraid?
Should the presence of a kind earthly father

awaken fears in any one, much less in his own children? I am your Maker and your Father, and you are my child—my very dear child; child by creation and also by redemption, therefore you need not fear." I became composed, but felt that God had selected a very poor instrument for so great a work.

Then the voice went on to say; "I have chosen you to be my second Christ; I have appointed Jesus, my son, to instruct you and make you wise in all things—to do my will in the great work of man's salvation." I remarked (for I began to have some confidence and assurance, the words seemed spoken so kindly), that the scriptures would not be fulfilled unless Jesus should come the second time in the clouds of heaven, and every eye behold him. To this the voice replied that I should begin preaching in a very humble manner, but through the wonderful wisdom and unction that should characterize my preaching I should become very noted. At the close of my career I should be standing in the pulpit of one of the best churches in the land, and while addressing the people in the most fervid manner and while all eyes were intently beholding me I should be caught up, passing through the roof without injury as the angels appointed would remove the timbers and with a mighty hand scatter every fragment of the church to the four winds. Continuing my upward flight I should be seated upon a cloud which would be prepared for that especial purpose, and there remain until the world should be made to revolve upon its axis until one entire revolution had been made, thus giving all the inhabitance of the earth a view of their Messiah or Savior.

After listening to this most (to me) remarkable declaration of my calling, future course, &c., I returned to my study and continued the sermon which I believed was being written by inspiration. I noticed, too, as I took my seat, that my inspirer, who called me brother, could speak to me as did the father and in the same "still small voice." From this time onward any remark he wished to make to me, instead of dictating the word which I wrote down, he addressed me directly —not apparently my mind through the instrumentality of my ear—but my mind directly.

I continued at my sermon but a short time when matters of a domestic nature engaged my attention and the sermon was laid aside for the time being.

CHAPTER IV.

It will be proper here to notice the person alluded to at the close of the communication in these words: "A man has been appointed to be your associate in this work; you will be his guide and instructor, and so together you will be thoroughly furnished unto every good word and work."

This man alluded to was my particular friend. Some five or six years previous to the time of our acquaintance he experienced religion. It was at a time when all his mental and physical powers had arrived at their fullest maturity—at a time when the excitement of his emotional nature would not be likely to lead him astray—but in the calmness of his mature judgment he would be fully competent to decide that he had met with that change denominated in scripture, "a being born of God." A short time after he had met with this great moral change, he felt deeply impressed that it was his duty to preach the gospel. For this holy work his previous education and employments, instead of being of that kind which would have filled his mind with stores of moral and spiritual truth, filled it with every thing but that kind of truth.

Still unfitted as he was, he felt a clear conviction that it was his duty to proclaim that truth which had emancipated his own soul from the bondage of sin, so he resolutely set himself to the work of preparing sermons. Not being able to draw very largely from books, and having so limited an acquaintance with the sacred Scriptures, he found that his work progressed but slowly. In the intervals of rest ad application he resorted to the habit of pacing to and fro in his room. This proved more successful in the matter of sermonizing than application to books, for at intervals strange and startling ideas came rushing into his mind. These he secured by jotting them down upon scraps of paper, and pasting them upon the wall of the room. After pacing thus for a length of time, he found that a sufficient number of ideas had been gathered together to form the basis of a good practical sermon.

Having met with such unexpected success in collecting ideas, he resolved to continue the practice in the future, rather than pour over musty volumes of theology to obtain them. But what might seem somewhat remarkable to some, he found, by often repeating this method of preparing sermons, that the ideas came more and more frequent and with greater distinctness into his mind. Thus he continued until after the lapse of a considerable time he found that pacing the room to and fro was entirely unnecessary. By placing the text at the head of the page, then composing his mind to a quiet, passive state, he found that ideas flowed into his brain as fast as the hand could place them upon paper. After the expiration of two or three hours he would find, upon reading, that a well arranged and very appropriately illustrated sermon had been penned.

At this time he knew not the source of this strange kind of inspiration. He supposed, as he expressed to me, that it was simply a gift—a rich legacy from his creator.

He invited me to his study for the purpose of witnessing some displays of his wonderful powers of composition. We entered his study together, and being seated he requested me to select any subject I might choose and write it down at the head of a sheet of paper, and he would proceed to enlarge upon and develope the subject without any premeditation whatever. This I found he could do with the greatest ease, and with very great rapidity; but what seemed to me the strangest of all peculiarities about this manner of writing was that he knew nothing more about what he had written down than I did myself before reading it. All he could say in answer to questions propounded was that he rendered his mind as perfectly passive as possible. Thus as one after another the words came into his mind, as if dictated by some invisible person, he wrote them down with the greatest possible dispatch. At the closing up of the writing upon each subject given him he seemed somewhat physically exhausted, but his mind showed no signs of languor or fatigue.

After repeated experiments in attempting to determine the cause of this mental phenomena, I came to the conclusion that a subject of so grave a character could not be pronounced upon with the limited number of facts before me.

To my clerical friend the facts presented were as inexplicable as to myself. Had he been controlled physically—that is, his hand and arm been made to move mechanically and without the volition of the mind, and the ideas written down without the knowledge of the subject—I could have come more readily to the conclusion that it originated from invisible intelligences. But that words could be dictated to the mind of the subject and by him communicated to paper from an invisible being was vastly more difficult to comprehend.

I repeatedly asked him in regard to his not understanding the matter of the subject written upon until it was read to him. He seemed positive that he could not bring to mind a single idea until he had read the subject matter of each communication. I remarked that it certainly was a phenomena of mind which no writer upon that subject had ever alluded to, and with this remark we parted for the evening.

A few days after this interview with my clerical friend, the thought occurred to me that I could bring a test which would more fully determine whether the ideas originated from some hitherto latent powers of his mind, or from a spirit intelligence.

The test which I thought of proposing was to present to him a subject to be written upon which I was satisfied that neither of us could possibly know any thing about. The subject which I thought to propose was that of the natural birth—a subject which all must admit is shrouded in great mystery, especially as it regards the origin of the human soul, as to whether it is from either parent or directly from God. I here subjoin the question and the reply given to it:

QUESTION.—How is the human soul introduced into the unborn child, and from whence do the souls of men come?

ANSWER.—There are two realms—first, matter filled with spirit, and second, spirit free of material incumbrance. The great spirit world in which all the material worlds are floating is spirit positive; the soul world is spirit negative. In this tremendous ocean of spirit positive there floats myriads of existences in the form of Monads, rained down from the Infinite, each one a particle of soul given off (so to speak) from the Great Eternal Mind. The Monads are not spirit negative, such as is contained in and constitutes the world of matter in all its million forms of beast, bird, reptile, and vegitation. Neither is it spirit positive, such as constitutes the sea, in which the worlds do float and whose purer sphere constitutes the breathing element of disembodied spirit,—but they are the original soul germs of immortal beings. They are particles of the *Deific Mind*—unique, unpartialed, homogeneous, old as Deity yet young as the new born infant—always existed, ever will exist. If you could see as disembodied spirits see, your gaze would be directed to the fine, soft, and glorious silvery sea spreading away in all directions. In this clear expanse floats uncounted myriads of these globular, unparticled monads, all infinitesimal in volume and each one enveloped in a fine electical substance which surrounds them perfectly. The spiritual waves bear them on towards the earth life. By a mysterious power they are drawn to the human male brain through the lungs. There they remain until a certain physical act is performed, when they descend, effect their mission, and the incarnation of a soul is effected and made ready to enter upon a new career of being.

After reading this reply to my question,

we both agreed that the subject was admirably treated and, so far as *we knew*, very truthfully; for it seemed but an expansion of Solomon's declaration: "The spirit shall return to God who gave it." Thus matters stood when the phenomena in relation to my own mental exercises occurred. The experiences of my friend were so very nearly similar to my own that I could scarcely have a doubt of his inspiration. Only these differences seemed manifest between his experience and my own, his inspiration came by degrees—my own came suddenly. His mind, like my own, was impressed with the words, but not, as in my case, so distinctly as to appear to the mind as words spoken by another.

The reader will remember that the spirit intelligence that addressed me in the garden claimed to be the Almighty, but the one who dictated the sermon represented himself as being the Lord Jesus Christ. Notwithstanding their positive declarations as to their identity, and all the evidence corroborative of the fact, an occasional doubt would flit across my mind. I knew the utter folly of obeying the voice of an invisible intelligence unless there was sufficient evidence of the identity of the persons who spoke, and as the awe first inspired by the alledged divine presence had begun to subside I, therefore, resolved to ask some questions which, if answered, would resolve all my doubts and make my duty plain before me. I was willing to obey the Lord in any thing He might require, if I could really satisfy myself that it was the Lord Jehovah who had called me. I therefore addressed the intelligence in language somewhat after the following manner: "If thou art the Almighty Father of us all wilt thou inform me as to the manner in which the worlds were made and placed in the positions they now occupy."

To which the intelligence replied in manner something like this: "My son, I will answer your question because it would not be right to require a child to perform a work of so grave a character without full and satisfactory evidence of the divine authority of the command. I will, therefore, to further assure your mind, and to resolve all doubts of my identity, give you a condensed history of my work as the Creator of the universe, or rather of the seven universes which I have already made."

CHAPTER IV.

"In revealing this to you, my child, I shall use familiar language—it would not become me as your maker to use other.

"Know, then, the gradations of matter or substances: First, *Spirit* is the inner principle of matter, mind, or soul, the inmost of spirit; and *Think* the life-principle of soul, and God, your Creator, the inner life-giving principle of think; therefore I am the All in All. The vast ocean of spirit essence in which all the worlds and universes of worlds are floating, is spirit positive, and it is the inner life-principle of matter in all its varied forms of crystalization, plants, trees, and animal life.

"The material worlds, which are the source of all spirit existences, are made from the great ocean of molten matter which forms the basis of this universe of worlds. From this vast, and by mortals incomprehensible, ocean of liquid fire there arises clouds of vapor, or matter in the sublimated form. These clouds of matter are set in motion by positive and negative currents of electricity. The friction caused by the rotating motion produces condensation, and condensation evolves heat —the heat causes the mass to retain the fluid condition until the increased velocity of the motion causes the immense mass to throw off large bodies of the molten fluid, these form the primary planets of the system. They in turn, from the increasing rapidity of their rotating motion, first assume the oblate spheroid, and then from their equatorial surface other masses are evolved which, after assuming the globular form, constitute the moons or secondary planets.

"But when the elemental matter of the primary planets, as it waxes more and more inflexible, is thrown off it often remains unbroken. When this takes place a ringed satellite is formed, and when the velocity of the planet is still farther increased, and the amount of matter remaining in the vast orb is sufficient, then two or three ringed satellites, besides several globular ones, are the result. The planet known in your world by the name of Saturn is an example of the kind alluded to.

"When the molten globe thrown off from the sun is of insufficient size to evolve the ringed or globular satellites, then it continues to revolve without them, as does your planet Mercury, without even a single moon to beautify the nights and render them more pleasant to the myriads who may inhabit it. The molten globe remaining after the matter constituting the primary and secondary planets has been thrown off becomes the central sun of that system. This sun, according to the laws which I have established, must contain in weight the exact amount of matter that all the planets and satellites put together contain which have been evolved or thrown from its surface.

"The youth period of each planet, embraces the time which elapses between the evolution or birth period, and the stage of the planet when its surface crust has arrived at a considerable degree of thickness, and the action of the elements has pulverized its rocks into the coarsest soils capable of sustaining the lowest forms of vegetation.

"The time which elapses from the first appearance of vegetation upon the planet until sentient existence in its myriad forms of animalcula, reptile, insect, and fully perfected animal life has made their appearance, is called the manhood of the planet.

"But as there was a time when the planet, by reason of its heat, could not sustain vegetable or animal life, so in the lapse of ages there must arrive a time when the earth shall have parted with so much of its heat that its frigid and desolate condition would not admit of animal or vegetable life being produced even in their lowest forms. This period is called the old age of the planet.

"Thus you must perceive that worlds wear out, waxing old us doth a garment. At this stage of the planet's existence the law which sustains them in mid-heaven is suspended, and they drop, like untimely fruit, into the bosom of mother ocean from whence they sprang. Having been dissolved by the fervid heat of that vast ocean of liquid fire, they are again raised in vapor, to go through the same process as before, of nebulous formation, condensation, and new world-formation.

"You will therefore perceive the grand and awful truth that with every world, as well as with every rational creature, there must arrive a period of time called the judgment day. Jesus, my Son, informed you that in the last day he would come in the clouds of heaven, with power and great glory, and then shall he sit upon the throne of his glory, and before him shall be gathered all the nations, and he will separate them as a shepherd divides his sheep from the goats, and he shall set the sheep on his right hand but the goats on his left. Then shall he say to those on his right hand, 'Come, ye blessed of my Father, inherit the kingdom prepared for you from the foundations of the world.' But to those on his left hand will he say, 'Depart from me, ye accursed, into everlasting fire, prepared for the devil and his angels.' You will perceive from this, my son, that not only the worn out worlds are cast into the lake of fire, but all the wicked—the chaff, the unfruitful trees—go with them. But the children of the kingdom—the wheat—shall be gathered into my garner above, to go no more out forever.

"But you will ask to what garner are they gathered. I will tell you. Very far above the universe to which your planetary system belongs, there is another universe of worlds, made up of infinitesimal particles thrown off from the worlds of this universe. These finer particles ascend as they escape from each world just on the same principle that vapor ascends from the ocean and forms clouds; or on the same principle and in the same manner as the grosser elements, from which your worlds were made, arise from the great ocean of unquenchable fire below.

"These infinitesimal particles are aggregated and condensed into globes of immense size, from which are thrown off their moons and ringed satellites, just as is done in your grosser material universe, only that this whole universe of worlds is beyond all comprehension of finite creatures most grand and inexpressibly glorious.

"The fine, soft, ethereal sea in which these countless suns and worlds are floating, is the source of a life of as much higher type as human life in your world is above that of the snail that crawls with measured slowness. But grand and beautiful as those worlds may seem to be, as compared with the worlds of your own universe, still there is another universe above that still more grand and glorious. The material of this universe is derived from the one below it—being composed of the inconceivable fine, soft, and glowing particles which rise like a vast exhalation to the regions above this universe, forming, when aggregated and condensed, into worlds by similar laws and in the manner alluded to before—another universe, as much more grand, and imposing, and beautiful as the one next below it is more glorious than the rudimentary universe to which your sun and planets belong.

"As the first universe above your own is called the first heaven, so the one just described by the order must be the second; and another above that, made up of rejected particles from the second, is called the fourth universe but the third heaven; so another above the third having been created in the same manner and by the same general laws which I have established, is denominated the fourth heaven. Another above this called the fifth, and still another called the sixth heaven, when you reach the universe globe called the seventh heaven, or the heaven of heavens. Of the glories which attach to and surround that super-celestial abode of the righteous no finite mind has or can conceive.

"If the light from your sun clothes all nature in your world with such glories as you behold, what, think you, must be the transcendent glories of that light which emanates from my great white throne, and these particles of the finest, purest, and softest light, reflected and refracted from countless forms of immeasurably refined plants, flowers, shrubs, and trees?

"On this vast universe globe which constitutes the super-celestial heaven, or the heaven of heavens, is to be found the New Jerusalem. Its walls are represented as being built of what in your world is considered the most precious and costly stones; the magnificent and lofty gates of solid pearl; the streets paved with gold. In that great city is to be found my palace, and within that the great white throne, which is the throne above all thrones, principalities, and dominions. There my glory dwells, and there centers all that is ineffably sublime and exquisitly unique and beautiful.

"This magnificent city of all cities constitutes the audience chamber of the great, grand universe of universes, or heaven of heavens. It is the place where I delight to honor those who, under great tribulation and in the fiery furnace of severe trial, have honored me.

"But grand and magnificent as all these vast universes of worlds may appear to you, they constitute but a very small portion of my vast empire, for outside of these universes of worlds and upon every side is to be found my *chaos*, or matter in an unorganized and unparticled state. But my empire does not stop at this point, for the infinite regions beyond this I call my space, where through the ages to come I intend to display my creative power to admiring universes of intelligences in the production of still greater and more magnificent universes of worlds than has hitherto been beheld. This shows me to be a being that delights not in inglorious ease, but in working out the great, grand problems of universal empire.

"But do you suppose that the organized universes of worlds which fill up with shining orbs the different heavens and the chaotic universe and the universe of space to be the full extent of my empire? I tell you it is not, for my empire extends as far beyond the utmost verge of my space as it is possible for me to think; yea, it extends farther than that —it extends even as far as I ever will think. Thus, if you can conceive of so vast an empire and not let reason totter from its throne, you have a faint outline of my past and prospective handiwork.

"Men of your earth have thoughts, and when fully matured they seek to externalize them in all the varied forms of mechanism, sculpture, painting, poetry, prose, and the like. So the created universes and the mighty hosts thereof are my thoughts externalized. These thoughts proceed from me in regular order, and constitute the grand *procession* of *the ages.*

"As insignificant as man may appear to be in his unregenerate, unrenewed state, still when born and quickened by my spirit and created anew after the image of Jesus, my son, he becomes a microcasm—a universe within himself. An angel might comprehend one single faculty of man as it now is, but to comprehend that faculty as it will be in its full, perfected strength, the mightiest angel intellect would be inadequate. Could you know what man must become to be the master-piece—the crowning glory of the infinite uncreated mind—you would then know in what estimation man's powers of mind and soul should be held. How very little man knows of himself. There yet remains immense oceans, gulfs, bays, and even whole continents of yet unexplored territory within that sublime and wonderful composition, *The Human Mind.*

"But know thou the order and gradation of spirit being. The unsaved of earth as they drop their earthly material dwellings pass into the air and are invisible to those remaining in the form. So the saved of earth life, when their spirits are freed from the mortal prison, pass up and beyond the air to the first heaven, but so very much finer and purer are these spirit forms that they are entirely invisible to the gross and impure spirits of the air. Also those of earth life who having become so morally pure as to be prepared to enter the second heaven are, as they escape from their bodies, invisible not only to the grossly impure spirits of the air, but also to those in the first heaven. So also those who, by right living and right doing have fitted themselves for an entrance into the third heaven, are also so etherial and refined as to be invisible to any below them in the heavenly scale of being.

"This same order continues until the highest, the super-celestial, heaven is reached. There the only invisible beings are myself and my son, who is equal with me in every divine perfection. But as we can reveal ourselves at pleasure to any and all the creatures we have made, so by special permission the intelligences of each of the several heavens may reveal themselves to those below them in the divine order of gradation. This order we have established that the higher orders of intelligences may be the better able to rule those below them.

"By this arrangement intelligences from all the different heavens may visit your earth, and yet the corrupt, wicked spirits of your air would be as insensible of their presence as those still remaining in the body.

"As the Anglo-Saxon and German races have now grown so sensitive that spirits can impress their minds, and in many instances control their bodies, so I have determined to introduce a new dispensation, and for this purpose I have called you and your friend to open and prepare the way for its introduction. Be humble, prayerful, and faithful, and all will be well.

CHAPTER VI.

This Communication, with the one preceding it, so won on my confidence that I scarcely entertained a doubt but that this emanated from the source it professed to. I felt that it must be a stupendous piece of folly, or a magnificent scheme for the glory of

God in the salvation of all the nations of the earth.

I often visited my friend, and as I then believed my divinely appointed companion in labors for the introduction and organization of a new church whose ministers should all be inspired and imbued with the highest wisdom, and that, too, direct from the great author and source of all wisdom. We conversed together upon all these great topics thus far introduced, and we were of one mind. Our moral sympathies seemed to flow together like two streams of water.

About this time *as we* were conversing upon the all absorbing topics of the new dispensation (we were sitting in his parlor) all at once his countenance assumed an altered appearance,—the tone of his voice changed, and I soon found myself conversing with another person. He was in what I considered the *trance* state. The spirit intelligence controlling him claimed to be his twin sister. The conversation turned upon the subject of her decease, and especially the manner in which she had succeeded in gaining first mental, and then, as at present, entire physical control. She stated that she was fully aware of her brother's unfitness for the gospel ministry because of his early training, and his great ignorance of the Scriptures, and theology in general. But such was her sympathy for him and his family that she had determined to overcome all these obstacles by obtaining from lofty intelligences such aid as he might need in the preparation of his sermons. She also stated that she was very much endeared to her brother, being an only sister, and she felt great interest in the great enterprise we were about to engage in, and assured me that we would have all necessary aid, and particularly that her brother could, in time, be controlled by spirits so as to astonish the world by his divine eloquence and words of wisdom.

She desired me to aid her brother in filling the position he then occupied as under shepherd of the chosen of God. She also directed me to purchase a blank book, and to carefully copy into it all the communications they might from time to time give us to properly prepare our minds for the great work.

I wished to be informed how soon we should be ready to commence our labors. She replied that it would be some considerable time before the great ship of the new dispensation would be properly freighted, but that as soon as all things were ready we should be notified. She then went on to state that we might expect, through her brother, a communication from some lofty intelligence upon the present aspects of spiritualism, and why it had made its appearance in just the manner it had. From this time onward she stated that if I wished to converse with her personally at any time I should signify it by a distinct mental desire, and she would immediately cause him to pass into the trance state, by gaining mesmeric control of his vocal organs, and then would answer any questions I might propound. With this arrangement she released him from the condition he was then in, and immediately he remarked as he gained control of himself, " I think I must have been asleep. " I informed him that he had been in a trance state and had been conversing for some time. He wished to know what he had said. I informed him of some things that we had conversed about, but some I kept to myself.

CHAPTTR VII.

I here subjoin the communication received through him, as alluded to above:

COMMUNICATION.

"How strange that men always are anxious to learn that which is perfectly useless. Angels and spirits do not attach any importance to many of the propositions which your teachers of theology are so anxious to sustain, because it matters not at all whether we adopt the theory that this globe was, from its infancy to its present state, progressed out of chaos by separate acts of creation under the fiat of the Almighty, or whether we believe the progress of growth has been one of development out of the life principle so imposed upon the new world at its birth that time could not go on without their unfolding gradually and according to a law.

"The great fact is admitted by all, independent of these theories of growth, that there has been what may be conveniently called *Creative Epochs* in this world's history, which are distinctly marked as divisions of time, though the precise beginings or endings have eluded the search of the best of our sciences. There was a time we know, as do you, when your earth, now so beautifully clothed with vegetation, was bare of all growing things. So there was a time when that vegetation began to creep over this earth's surface. There was a time when animal life could not exist by breathing this atmosphere, therefore there must have been a time when animal life had a beginning. There was a time, too, when man *was not* and when he began to people the earth. All of earth is not eternal, because it is not immortal; not being immortal it must have had a begining, and that, too, by a law older than itself. These epochs have come gradually, not only in reference to the whole earth's great development, but this is not all. Each epoch has in itself been the subject of a gradual introduction and growth, and a gradual decay and disappearance as it has given way to its succeeding epoch.

"Remember that each epoch is but the foundation for or on which the epoch succeeding it has been built *up*. Each epoch has sprung into being not complete and finished, but from germinal beginnings that have found their life and sustenance in the ashes of the past—each successive epoch furnishing in its ashes material for a higher growth in the scale of being. These epochs have succeeded in regular series, and the last so-called act of creation was the coming of *Man*.

"Far back in the East we discern glimmerings of light upon the question; when and how the human race began its career on the earth. But they are merest glimmerings, and convey to you nor us no more than the beautiful, simple, record of the Bible that God created man in his own image and called him Adam. *But mark* through what vicisitudes of life—what changes and varieties of condition. What growth and refinement, physical and spiritual, this race has been brought to its present fine development can not be stated in any brief compendium.

"That this world, however, is progressing as heretofore to some mighty yet higher condition, and that the beings that are ultimately to inhabit it will rank higher in the scale of being than its present inhabitants, is incontrovertibly inferred by all analogy, and is received by all Christians at least—if not by all civilized people—as an event which awaits only the sure fulfillment of prophecy. No wise human being will dare to say that even in his life time there may not be developments promising things yet to be which were never dreamed of before his day. You know not where to look for the coming great change. But we know, through the Christian dispensation, what the signs shall be when the great change approaches. That it will be gradual we infer from analogy. That it will come silently, without proclamation, like a thief in the night, we know, and you *believe*, from revelation.

"It is but a few years since the American nation was surprised and amused with the tidings of what were first called the Rochester knockings. By most persons the story was entirely disbelieved and deemed totally unworthy a second thought, much less of serious consideration. *But mark*, from that little beginning what a strange progress and startling development the thing called spiritualism has made. Subject to ridicule the most sarcastic that could be invented; to instigations and tests as various as there are varieties of conceits in the human brain; explained over and over again by as many different theories as men had time to examine —theories frequently militating against each other, so that the defender of the cause can often find his best arguments in the mouths of those who think to condemn it—the most educated class of the people all arrayed in stern opposition; the church thundering its anathemas against it, with a bitterness, too, which, had it been sustained by public opinion, would have brought the early votaries of spiritualism to the fiery stake.

"Little understood,—often misunderstood and condemned,—used and abused in every way—still the glaring fact remains, that no cause, moral or religious, civil or intellectual, physical or spiritual, ever made such progress in securing the more or less enlightened faith of men than this same cause of spiritualism.

Its active opponents seem to have given up the fruitless attempt to stop it, and to have sunk back from their labors, seeking consolation in the thought that if it contained not truth it could not prevail. They have left it where indeed they have found it, in God's hands, to manage according to his own wisdom and high behest. As to the result, as far as our observation goes, the community is divided into two large classes, namely—those who believe in spiritualism in the broad acceptation of the term, and those who do not believe in it, but think there must or may be something in it. The number of those who utterly reject all its facts and phenomena as trickery are too small to be named as a class.

"In order that you may become a mighty instrumentality in the hands of our Heavenly Father, we are permitted to point out the *rocks on which thousands of poor mortal's barques are wrecked through the abuse of spiritualism*—through ignorance of its laws and a lack of knowledge of its vast powers.

"My brother, pray heartily for God's great blessing on your work and ours,—pray that there may soon come more positive fruits of our united labors; and when you and we, in the grand future, shall have to confess the wonder-working of an all wise father who rules these great events as he does all others —yes, even the confession of your by-and-by deep experience that shall awaken your spirit into new life, leading it to pray daily that it may be so privileged of God as to do its humble part in bringing his kingdom on the earth, in seeing that his will is done here even as it is done in heaven.

"Reverently we write what we do know and what we have received from others who can not lie. We would be humble even as a little child seeking for truth with God's blessing on our prayers and efforts. My brother, let me say to you that a large proportion of those who believe in the actuality of the phenomena have no confidence in it 'because they see neither in the *mediums* nor in much of the *phenomena* any *special characteristics* that mark the *high* or the *holy*, for alas, they partake of all degrees, from the highest of heaven's blessed truths to the lowest of hell's horrors.'

"The wrong term has been used—it should have been spiritism, or demonism in the original sense of that term, because if the good spirits can come to bless why could not the bad ones come also? If the low could come, why not the high and the pure, seeing that the same law governs both worlds? God works by general laws and special providence.

"The thinking world finds itself continually perplexed, sometimes beyond endurance, by the absurdities, the contradictions, the follies, and even the wickedness, that breaks in upon them under the guise of spiritualism. Alas, poor humanity, weary and worn with its laborious search after truth—often ready to sink in the turmoil of doubts that surround them. They welcome any explanation, incomplete though they know it be, as sufficient to afford them a retreat wherein they may have some rest. They do not intend to deny the fact of the manifestations. They know that they can not be denied and so give their willing consent to the scientific theory of *odilic force*. The world was satisfied for a time (being weary) to rest within its shelter. It sufficed to give them a moment's respite only to renew the inquiry with increased earnestness, determined by God's aid, to find the truth which they knew only awaited those who strove humbly after it. 'Knock and it shall be opened unto you, seek and ye shall find' were blessed words of encouragement which gave them new strength. They seek the truth only for the truth's sake. They trusted that God would guide them through all their deviations from the right path. This is the state of the truly pious world on your earth to-day—yes, of millions. They are praying to God that if there is any good in these manifestations that they may learn to know it in their own experience. They ask to know in their own consciousness the external and internal cause of the presence of the spirit world about them, and glory be to God, at last the answer begins to come.

"Thousands are becoming conscious of some power communicating mighty and mysterious influences. Each day is advancing the truth. Thousands scattered up and down the highways of life feel a power mysteriously aiding their search after truth and saving them from errors committed by those who have not known the wonder workings of a true appeal to the Great Father of all spirits. Yes, and think you not that a new organization and church government is not necessary; think you not that new ministers of a mighty and glorious revelation is not demanded by this power?

"In this, as in all other subjects that may interest the human mind, too much or too sudden knowledge topples the reason and opens the way for folly to enter in. God in his mercy has sent the world this message— 'Ye shall have the truth as fast as ye can bear it, for if it should come as fast as it could be given it would craze every brain.

"We have thus shown you of the world's progress through certain stages of development in spiritualism. By this you will see that while such forms of mediumship have their use for the purpose of introducing spiritualism to the world's notice, they are far from the highest forms. The highest form of mediumship is that in which the individuality of the medium is most developed and the most active, so that the medium self being a spirit in the body may draw directly from the great spiritual fountain of God's infinite mercy, truth and power. In other words, the highest mediumship is what has hitherto been vaguely known as *inspiration*, and the desire and capacity for it depends upon holiness of heart in the individual; but each one must work out his own position in very truth, but not without aid. The power of prayer is mighty. The Great Father of spirits will send us such influences as we truly need and ask for. 'Ask and ye shall receive,' yes even the desired presence of the blessed spirit of Jesus.

"True spiritualism (not spirit-ism) teaches us that truth is handed down by gradation from the great central fountain of God's eternal wisdom and love through the various conditions of spirit life which rise from one to another in the many mansions, each nearer to the source of direct inspiration—the all-wise Father, even the Almighty.

"I would have you know that a mighty struggle for dominion between the powers of darkness and the powers who love the good is now going on, and that it is yet to sweep over the entire globe, for the wisdom of God is even more mighty than the folly of evil.

"It is also an important consideration that your attention will be directed to the more spiritual development of spiritualism which now begins to be gradually unfolded. It must be seen that an equal development of head and heart are necessary to make the perfect man. We believe the *heart must first be cultivated, must receive the divine annointing* or the head cannot receive true wisdom. Except the heart be right, the head will be full of errors, that lead the spirit to its ruin.

"This is no new proposition; the philosophy of it is simply this: True heart development only brings that peace of mind which fits it for the higher intellectual conceptions, fits it for the reception of the highest truth. The nations who boast of the highest cultivation have ignored it altogether and set up intellectual idols.

"Wonderful has been the intellectual, and as a consequence the material progress of the nations, and particularly during the past century. But is it not true that spiritual culture and development have been retrogading? Witness the sad results, see the utter selfishment of the trading classes. With a few noble exceptions every man of them is striving with his whole soul to find out not how he can help his neighbor, but how advance before him or overreach him. Alas for such Christianity. We verily believe that the founder of their religion would hardly recognize in them the faintest lineaments of of discipleship.

"The feeling that true spiritualism should have something to do with the understanding of the heart or the moral feelings, and the fact that the present manifestations of it so far has exhibited very little, if any, is one cause of its greatest opposition.

"In conclusion, let me say to you again, my brother, be humble—humble as a child before God. Be ready when the time shall have arrived to commence the good work—rejoice, for you will find the yoke easy and the burden light.

"Be assured that to seek spiritual knowledge for trifling curiosity is to expose yourself to the penalty of sacrilege. By the fixed law of heaven you will first get what you seek after. Beware lest you bring to us too careles a heart or a head too vain of its understanding. Do not suppose that you can turn away and neglect these things for any motive with impunity; your likes and dislikes can not change the orderings of God's providence. Make yourselves what you know you ought to be, and you will learn to thank God for the sweet angel influence that shall guide and guard you through every hour of your life. If God's holy angels can and do come, why may not the blessed spirit of Jesus come too? Has he not come already; is he not in the midst even now and we know him not?"

CHAPTER VIII.

Now, kind reader, suppose you had received such a communication just upon the close of the previous wonderful announcements?—a communication containing, at least to me, much of knowledge, interspersed with cogent reasonings and pious persuasions, cautions and exhortations. I say what would you have thought of it? How would you have acted or how sustained yourself? Would not your reason have been overpowered by the accumulation of evidences after evidences that all these wonderful aperations must be of God? How could a man suppose that beings ever existed who could practice such deceptions upon a man who had but one purpose, and that purpose to do the will of God?

Let us recapitulate a little. First. Here is the experience of a minister, for years, whose every sermon was either in its sentiments or in its very words so distinctly impressed upon his mind, that it amounted to actual inspiration. Then, not like my brother who through years was brought into this condition, I was suddenly brought into the same or a much higher condition of inspiration, apparently by the same power. I say higher, because my mind became so sensitive that not only the very words of the spirit intelligences were heard, but also the intonatiions of their voice, as well as style and manner of expression was clearly perceived. Then, added to this, the wonderful call of what purported to be Jesus himself, and this very wonderful call indorsed by one who styled himself the Almighty—who declares that he intended to make me the great apostle even the very Christ of a new dispensation—and then to cap the climax of the strange and the marvelous, this very intelligence, who gives his name as the Almighty God, to further gain my confidence, gives to me what might be called a detailed summary of the origin and formation of the myriad worlds belonging to our universe, and of all the infinitely refined worlds which belong to the universes of the different heavens, besides many other wonderful matters pertaining to the final condition of the wicked and the righteous; and finally in the last subject presented it announces the fact that angels and spirits of just men do not place the same estimation upon theological data as the children of men in the form do. It announces the order of God's procedure in the production of this world, and that all the various orders of plants and animals sprang from germinal beginnings, and from this is inferred the order of introduction of a now higher and more glorious dispensation of the Christian religion!

Reference is made to the "Rochester Knockings," and from that, as a germinal begining, has sprung the whole entire system of spiritual manifestations and spiritual philosophy, that spiritualism had its glaring defects, being addressed to the senses and intellect and not to the moral nature of man. But that this very feature was intentional on the part of God, as the head and heart condition of the world would entirely preclude any other kind of introduction, that God had sent the world his message: "Ye shall have the truth as fast as ye can bear it, for if it should come as fast as it could be given it would craze every brain." Then it closes with a very solemn exhortation, with many cautions and instructions as to how we should begin the work. The yoke of God's service is promised to be made very light, and the very gratifying intelligence announced that angels and the blessed spirit of Jesus was present.

Now, kind reader, in view of all these announcements, declarations, and instructions, think you that your judgment would not have arrived at the same conclusion which my own arrived at, to-wit: that it must be of God, and therefore a grand, magnificent reality? It will be noticed that in this last communication, through my friend and to be my fellow-laborer, all my doubts respecting the present morally defective aspect of spiritualism, were solved as by a magic wand, for that defective aspect is declared to be the divine arrangement, in order to save the world from being crazed by the sudden introduction of such glorious light.

But, I must here notice my conceptions of what would be necessary in order to insure the complete success of a new and more glorious dispensation.

First, it seemed necessary, that a perfect system of theology should be announced, and rendered so clear and obvious that all the glaring defects in the various systems of faith received in the different branches of the Christian Church might be made to appear, which would lead to their speedy removal.

Second, that the truth revealed should be accompanied by greater and more vigorous displays of the divine power. Inasmuch as revelation, so far as it *has been given*, has been given *progressively*, that is, each succeeding step in the process being in advance of the preceeding one, we may expect that each of the succeeding series of revelations would be given by advancing the incoming one to a point far above the preceding.

For instance, as the Mosaic dispensation of

the church was a great advance upon the patriarchal which preceded it, and as the Christian dispensation was in every way more glorious than the Mosaic, so, in view of these facts, we might reasonable expect that a new dispensation of the church would in every respect far transcend the Christian; and this advance in a new dispensation I had fixed in my own mind to consist in separating the *truth* from every vestige of error, or in other words, to give a perfect standard of truth by which all creeds, and all systems of theology, might be tested. Then, added to this, I conceived that a much greater degree of Holy Spirit power was necessary, in order that the present Christian church might be baptized into a vastly higher degree of spiritual life and activity; in other words, that the incoming new and more glorious dispensation then must be placed upon the revelator's *white horse,* the emblem of purity and strength, with the bow of truth in its hand, riding forth conquering and to conquer, which undoubtedly signifies great success. With these ideas of what was necessary in a new and more glorious dispensation, I asked for a perfect system of theology, and a better method of organizing the moral element, so as to result in a transcendently higher unfolding of moral power, and this was in substance the answer:

CHAPTER IX.

A perfect system of theology has ever been the great desire of the poor, toiling millions of struggling humanity. They have looked to schools, to governments, to nations, to physics, *yea*, even to the sun, the moon, and the stars. But they have denied to mortals that which they sought, "a perfect system of divine truth, because they turned away from the angel guides within that urged and entreated man to worship God as the angels worship him, by realizing in his presence the joy which arises from a perfect obedience to all the laws under which they were created." Angels do not pretend to comprehend God's name, or his infinite power, but reverently bow before his throne to await the pleasure of his will, for to obey God in all things is to realize the highest fruition of which our moral and spirtual natures are susceptible. The Jews, as a nation, were guided by spiritual intelligences, as Paul says: "The law was ordained by angels in the hands of a mediator, and that mediator was Moses." After Moses other leaders were raised up who managed to keep them measurably united as a nation, until the more glorious dispensation of Christ, when they were dispersed, having answered the end for which they were kept together.

The great mistakes of the Christian world has consisted, in part, in not separating that which exclusively belonged to the patriarchs and Jews from that which is strictly Christian. When Jesus Christ, the beloved Nazarene, came, he set forth a system of truth eminently spiritual, which stood out in bold contrast with the grossly idolatrous systems of the pagan world, and the partially material system of the Jews. In order that his pure spiritualism might be propagated, he chose his disciples with great care, and instructed them in all that pertained especially to his system of truth in contradistinction to all others.

He bestowed upon them first spiritual powers, and instructed them how they might be cultivated. Did you possess the full record of his instructions, you would find that he directed them as to their modes of life, diet, moral and intellectual culture, every thing in fact that could favor the fullest developement of their several spiritual and physical powers.

His mission was to enjoin a pure, natural worship of God, even a worship in spirit and in truth—to tear down the old material shrines, bloody sacrificial forms, and to show that the truest devotion is that which is paid in daily practice which embodies Deity in the life of man, and makes every thought a secret power, a secret prayer, and every desire an aspiration. Such, dear brethren, should be your interpretations of your Savior's teachings, and the church, and the world will never be benefited until it is done, and all the followers of the Master have the moral courage to maintain this interpretation in their life and *inner being*.

Does not the holy record show that these spiritual doctrines were clearly understood by his disciples? If this be doubted, it is only necessary to refer to Christ's expressions when about to leave them. Said he, in a way peculiar to himself: "I am going to my Father's house, where I will prepare a place for you." This he said naturally, and it is folly to suppose that they did not so understand him. Then he assured them that they should not be left without a comforter, even the spirit of truth. This expression simply implies that the doctrines he had inculcated were inspired by the spirit of truth, which represents the Deity; for he says, distinctly and emphatically, that the spirit of truth inspired his every word. Yes, our brother Jesus fully and entirely embodied the spirit of truth, even the Holy Ghost, which should convince the world of sin, of righteousness, and of judgment.

The spiritual manifestions that accompanied his mission are worthy your closest observation, for upon them depends the whole power and beauty of his religion. 'Jesus taught a doctrine *higher* and *holier* than that of Moses, for without superceding the Commandments of the latter, he added to them a more important and comprehensive rule of morals, *The Law of Love*. From the lower classes in the community around him he chose his most intimate companions. To them he spoke in pure and simple language, unparabolic, without trope or metaphor. To them he confided the meaning of his inspiration; explained the principles of his religion. To them he defined the nature of his spiritual gifts, and promised them that still greater miracles should they do than he had done.

A thorough and impartial examination of the gospel histories can only lead you to the conclusion that Jesus Christ was the first to announce (that which constituted the central and paramount doctrine of his system) that the *purification* of the *individual soul* can be effected only by *repentance*, which is *godly sorrow for sin*, and by living a life of *self denial* and of individual effort for the good of others. This, and nothing else, can accomplish man's salvation. This rule in the least departed from reduces churches, households, individuals, in just the exact ratio of their departure from it. Earnestness was Jesus; special attribute and the secret of his wonderful power to influence others. So alone can it give you success in all spiritual matters, which you are to prosecute in the earth. He was sincere in all his convictions, and hence his power of stamping them upon the hearts and consciences of others. He was true to nature, and therefore true to nature's God. Hence his power to control her hidden forces in the cure of diseases and other miraculous manifestations. Jesus, in the example he presented to the world, designed in the fullest degree to display the possibilities of your actual *human nature*. Said he, addressing his disciples, greater works than these shall ye do. Therefore, it must be clear that it was the aim of Christ in all his teachings to establish a purely spiritual religion for *all men*—one not above their comprehensions or ability to do, but one in which all are endowed with the power to appreciate and fully understand.

I have said that the *spiritual manifestations* that accompanied the mission of Jesus are worthy of your *closest observation*. Take away from the New Testament all reference to spiritual manifestations and angel ministry; all teachings concerning the soul and its relation, such as are contained in the act of the transfiguration—the appearance of the angels in the sepulcher; of Jesus himself after the resurrection; the release of Paul and Silas and the like, and there will be left no real foundation for any religion whatever. There would be left nothing but the blind theories and vague conceptions of individual minds. Most emphatically the New Testament is a revealment of spirit intercourse or the laws and facts of spirit manifestations, and it is this fact which causes it to be the accepted religion of the world. If the spiritual element were extracted, it would be but a dead tree, a solemn mockery of words without any special meaning. Jesus was the king who was to come to Zion—the inspired prophet, sent of God, and you are to sit at his feet and learn of him.

The fact that Jesus held intercourse with spirit beings was fully believed by his disciples, and accepted as an established fact, and the disciples were accustomed to impart instruction as to the means of receiving spiritual enlightenment. Says Paul, walk not after the flesh, but after the spirit, for to be spiritually minded is life and joy and peace. And they also give warning against false inspiration, by false prophets, in this language: Try the spirits and if they shall confess Jesus, they are true, but if they confess not Jesus they are not of God. Even the apostle to the gentiles, while he engrafted on the new faith much that was absolute and even severe in the Hebrew Law, still retains the essence of spiritual religion, as when in his first epistle to the Corinthians, he gives a distinguished catalogue of spiritual gifts. Said he brethren I would not have you ignorant, &c., &c.

Now if these spiritual manifestations had been unknown to his co-temporaries, obviously Paul would not have thus so plainly described them, nor would he have put it on record that one star differeth from another star in glory. So men are variously endowed in these respects, and each man should be content with whatever gift his God had endowed him with.

These passages should show you what were the practices and beliefs of the primitive Christians, and they describe in very many respects, what is now called spiritual manifestations. While the great essential truths taught by brother Jesus remain, the doctrines erected upon them are widely different from the early and simple faith of the primitive Christians. So when the apostles had passed away, and the active, living, spiritual vitality of the church had disappeared, all spiritual manifestations went with them, and have not appeared again, until within a few years. The Roman Church still claims to be the only repository of spiritual gifts, but her frauds and fabrications have been too gross and palpable to enable her to sustain the character, and to-day, among the pure and humblest disciples of Christ, of any and all denominations, not one of these gifts are to be found. But notwithstanding these outward manifestations are absent, still Christianity has traversed many parts of the earth, and planted its banner on far distant shores, and where really known has shed a hallowed radiance upon every walk of life. Still it possesses not a tithe of the vital earnestness which characterized its founders.

This is because Christians regard doctrine, rather than practice, and follow abstract teachings rather than the living example of their great and noble founder. Because Christ is adored as an external, instead of a spiritual Savior. Because men prefer to bow before the cross, the emblem of vicarious suffering rather than take it up, sacrifising selfish ease and becoming a living, spiritual force in the earth. But blessed be God, vitality is not all departed. Surely the omnipotent spirit of truth *liveth still*. It survives all persecutions and even the chilling ordeal of neglect. In unlooked for forms and modes it comes, age after age, to unfold and record its constant lessons.

My brother, the great spiritual light of the present day, *imperfect* as it is impure, as are many of its advocates and teachers; polluted as it is by impostors, both in and out of the form, who drag down its sacred truths to the dust, and would render it subservient to every base desire; I say notwithstanding all these obstructions, it is to triumph and become a great and mighty river, of divine power, to beautify and fertilize the remotest parts of God's moral vinyard, and like a new star of Bethlehem, will guide the wanderers in darkness, and lead them to bow down at the shrine of truth. All men will learn to know that between the world of mankind and the realm of spirits, there is really no barrier— that all around you, in the haunts of solitude, in the retirements of your homes, in the crowded arena of active life, *invisible beings* watch over you, guarding your every footsteps, mingling their influence in every thought and every action.

The soul is endowed with many latent faculties. These are brought forth in many ways through the agency of invisible directors. Among these are the spiritual gifts, spoken of in the New Testament; the gift of healing of prophesy, of inspiration of tongues, and all other endowments spoken of in the Bible.

In conclusion, let me enjoin upon you to *mark well* the *fact*, that the foundation of spiritualism and spiritual truth, is as old as eternity; that it has been especially embodied in the life and character of the divine Jesus, and expressed in the Christian religion, and also in spirit influence *now being so awfully* and so grandly manifested, through you and through other mediums whom God has created for this purpose.

We have thus endeavored to give you a faint outline of a perfect system of theology. Be prayerful and patient; spiritual influences are around you preparing you for a vast field that will open before you soon. We shall never leave you for a moment, or cease our exertions in your behalf. We trust on your part, that you will not faint by the way, or suffer your spirits to falter, or your efforts to relax in the great and holy cause.

CHAPTER IX.

Thus, kind reader, you have gone through another, in some respects remarkable production from an invisible intelligence of an invisible world. Your attention has no doubt been called to the great leading fact of the composition—I allude to the fact of the full indorsement of Jesus Christ, and his pre-eminent system of self-denying love. That his embodied divinity is to be the sole divinity of all coming time; that his foundation of theological truth, being laid in the very nature and constitution of the human soul, must remain through the ages to come, as the true, altogether righteous foundation of all who would build thereon; that his divinity being indorsed by God himself, can not be set aside or overturned; that other foundations can no man lay than is laid, even the foundation of Jesus Christ the righteous.

As I have before said, I had concluded within my own mind, that in the introduction of a new system, Jesus Christ *must be indorsed.* I could not see how he could be set aside and his teachings set at naught. I had often heard spiritualists speak lightly of Christ, his teachings, and the Bible as an obsolete, worn out production of the past, and, like an old almanac, to be laid aside for something more new and altogether more wonderful. I had heard them go so far in opposition to Christ and his system of truth, that they could declare that all his miracles were an imposition, produced by understanding the laws by which mesmeric and psichological effects are produced upon other minds;

that when he declared to his disciples that he arose with his material body and was veritably flesh and bones and therefore could eat ordinary food, and did so eat, that he *deceived them*—he only having his spirit body so clothed with vapor and other material substances as being attracted around him would reflect the ordinary lamp and solar light, thus making his spirit body visible and the eating of food a pure deception practiced upon his disciples simply by mesmeric influences.

To all these declarations of spiritualists I had but one reply: Christ and his teachings must be received as the true inspiration of God or altogether rejected, as only proceeding from a base deceiver, who only received at the hands of the Jews, when crucified, his just deserts. To utterly reject him, I had conceived would be to blot the sun from the spiritual heavens; but to receive him as a true man, and make his teachings our ideal of rectitude, would be to place the spiritual sun where it ought to be placed in the spiritual horizon to give light to all.

I had thought that if one law governed all communications from the spirit world, that inspiration might proceed from an evil spirit as well as from one that was holy, and if so, then some test must be discovered to distinguish that inspiration which might proceed from evil spirits and that which might proceed from those holy and reliable. This test I found given by the apostle John, and in these words: "Beloved, believe not every spirit, but try the spirits whether they are of God, for every spirit that confesseth not Jesus Christ is not of God."

To apply this test to this communication would certainly result in receiving it, for though it indorsed modern spiritualism, it breathed the very spirit of the cross and most persistent loyalty to Jesus Christ. I was therefore compelled to indorse it as proceeding from God. But, though this communication appeared to breathe the true spirit of love to Jesus, yet the fact that it indorses, to some extent, the doctrine of modern spiritualism seemed to my mind a very great incongruity, for so-called modern spiritualism rests its claims to being received upon the declarations of so-called spirit friends and relatives; and if so, then, how could friends and relatives deceased but recently be expected to be able to solve the great questions of Christian theology? Could it be supposed that the simple transition from one state or condition of being to another would put the subject in possession of all knowledge? Of this I had many doubts—doubts arising from the very nature and constitution of the human soul, and from the positive declarations of scripture to the contrary.

The teachings of all known science upon the subject of the mind's growth and developement, so far as I had read, was to this import: The mind is like a seed which begins first with the germ, then the blade, and finally the ear fully perfected. So the mind begins to unfold, first by impressions made upon the senses; then, as impressions increase and ripen into knowledge, the blade of the mind appears; and so when knowledge is classified and generalized the reason fully appears. But the mind principle being immortal, it must ever continue to unfold, so there can be no time when it will be put in possession of all knowledge—death being but a laying off of the old garments of mortality and the putting on of the new and spiritual garments of immortality. I was, therefore, very naturally led to conclude that the human mind, after leaving the clay tenement, began and progressed in knowledge, wisdom, and activity just as it began its career in this state of existence. Then to consult them upon the great and all-important matters of the soul's salvation seemed to me to exhibit a lack of even a low degree of wisdom. The scriptures, also, as I had interpreted them, clearly taught the same gradual unfoldment of the human mind, for by them we are taught that the souls of the righteous at death pass into a place called, by Jesus, paradise—a kind of preparatory or intermediate state where they are further educated in the school of Christ, by being led to living fountains of knowledge; and finally, after the judgment ordeal shall have been passed, they will be awarded a place in that "house of many mansions," each taking his position according to the advancement he had previously made—not in holiness and moral purity, for we must know that nothing in the least impure or unclean can enter there at all—but in knowledge and wisdom, and moral and intellectual power.

But to return to the subject under discussion, viz: The objections to receiving the communications of spirits as the truths of God. I had said that I could not see the propriety of receiving them unless they fully indorsed Jesus Christ and his self-sacrificing love system, whether they purported to come from departed saints or angels.

So far as I had been made acquainted with the spirit world, through the different mediums, they bore no marks of any exalted wisdom, but on the contrary they seemed to proceed from beings in every respect like the great mass of wicked, ungodly men in the earth life, who hold all sorts of opinions upon every subject embraced within the range of human thought.

As the scriptures throw some light upon

the subject of where the wicked dead remain until the judgment, I will therefore present some of the evidences which are therein presented: Jesus said of Judas, that he went to "his place." If "his place" meant paradise then he would have used the same language that he used in reference to the thief on the cross: "This day shalt thou be with me in paradise." So he must have meant by the word "place" some other abode than paradise where himself and the penitent thief departed to on that day. In another place, Jesus, in speaking of the rich man, said, "that he died, and in hell; he lifted up his eyes, being in torments." Dr. Barns remarks on this passage that the word hell literally means an obscure place where the wicked dead are gathered after leaving the body. The new translation of Dr. Kendrick renders the word hell in this place the "under world." These two passages clearly speak of a "place" where the wicked dead have a conscious state of existence.

Now can we ascertain where that "place is?" Paul throws some additional light upon the subject. He says, "the saints in this life wrestle not with flesh and blood alone (that is, with the wicked in the body,) but with spirit wickedness in high places." As Dr. Kendrick renders it: "We wrestle against the spiritual powers of evil in the heavenly places or in the air." Paul renders it thus in the second chapter of the same book: "The prince of the powers of the air; the spirit that is now working in the sons of disobedience." By these two passages we have the exact location for wicked spirits and devils—the air of our earth. I believe commentators agree that the word spirits refers to disembodied men, and that devils refer to fallen angels. Hence the expressions, "seducing spirits, and doctrines of devils, possessed with an evil spirit, with unclean spirits, with a spirit of divination," &c. All these expressions indicate a difference between the spirits of fallen angels and those of fallen men. Paul states that he was caught up into paradise, or the third heaven. The first heaven, as the Jews understood the arrangements of the heavens, was where the fowls of air fly; beyond this, to the extent of the earth's atmosphere, the second; and where the sun, moon, and stars were, or the space beyond the earth's atmosphere, the third heaven.

Now these teachings of the scripture, in connection with many others, indicate very distinctly that the "*place*" to which Judas was sent to remain until the judgment, was where all wicked spirits are; with the prince of the power of the air, or in the air, which surrounds this earth. This being the case as I understand the teachings of scripture, I could very readily account for these spirit rappings, and manifestations, as coming from the departed wicked, or those who are not in scripture called saints. Now if a revelation were to be made from such beings, I think it would be characterized by every possible expression of opposite opinion on every subject they were conversant with in the earth life. They would carry with them their moral tastes, their likes and dislikes their habits of mind and their earthly knowledge. We might reasonably suppose, then, from the nature and constitution of the human mind, that if a revelation were made from them it would be characterized just as these spirit manifestations are characterized, by a revelation of contradictory opinions upon every subject that had occupied the mind of wicked men in the earth life.

Thus I accounted for all the contradictory statements, and opposition to Christ and his doctrines made by spirits through mediums. But then again, I arrived at what seemed to me a reasonable conclusion, that if the mediums were pure beings themselves, and they could be controlled by holy spirits, or by Christ himself, they would declare the same doctrines that Christ taught, only more fully stated and unfolded; and that also through them, might be displayed the same miraculous manifestations that were displayed in Christ's day, only in greater powers.

I had also formed the opinion that the reasons that the same spirit of earnest piety and humble submission to God pervaded and characterized the entire writings of the Bible, was owing to the fact that the writers of the different books, wrote as they were moved to write, or speak by the Holy Spirit, and being holy, of course there would necessarily be an agreement in doctrine, in fact, and in moral tone. Daniel says he was inspired by the angel Gabriel. John, the revelator, says he was inspired by one of the old prophets, yet he is called an angel; and they were told to write in a book, what was given them to record. So I conceived that all that was necessary, for a true inspiration, was to be controlled by a *holy Spirit,* one who loved God and the Savior Jesus Christ; and the communication could not be otherwise than truthful, whether given to me by one or a hundred of the *Holy Ones,* for all holy beings must utter only, and ever the truth. They can not disagree, they can not *lie;* if they should falsify they would cease to be holy.

Now this last communication, as well as the first, purported to come from a spirit who loved Jesus, rather than from Jesus himself; yet, nevertheless, I could but receive it for it seemed to breathe forth the spirit of love to Jesus, and also to God as his Father. I could not, for a moment, suppose that an intelligence that could manifest so ardent a love for Jesus and his doctrine, could *deceive*—not even in the least. Besides the spirit, whoever it might be, seemed to understand the precise cause of the deficiencies of churches and their memberships.

He points out, in the first place, the great mistakes of the Christian world, which consists in confounding that which belonged to the Patriarchal and Jewish dispensations, with that which properly belongs to the pure spirituality of the Christian dispensation. This we all know has given rise to divisions, and *great differences* of *doctrine* and forms of church government. Then, after pointing out clearly and briefly the mission of the Lord Jesus Christ, he states in what true devotion consists, in these words: "The truest devotion is that which is paid in daily practice, which embodies Deity in the life of man, and makes every thought a secret power and a secret prayer, and every desire an aspiration."

This appears to me to be the embodied essence of true religion. Then he calls special attention to the spiritual manifestations that accompanied Christ's mission, and says upon them depend the power and the moral beauty of his system of religion. He says, also, that Jesus Christ was the first to announce that which constitutes the central, paramount doctrine of his system; that the purification of the individual soul can be effected only by *repentance for all sins,* and by making individual efforts and self-sacrifices for the good of others. This, and nothing else, can accomplish *man's* purification.

He states that the design of the spotless example of Christ was to display the highest possibilities of our actual human nature. It is also distinctly stated that the New Testament is a revealment of spiritual intercourse, or the laws and facts of spiritual manifestations, and that this is a fact which causes it to be the accepted religion of the world. He also speaks of true and false inspirations, and the test that should be applied to distinguish between them. Christianity, he says, has shed a hallowed radiance on every walk of life, still it possesses not the power of *vital earnestness.* This is so, because Christians regard doctrines rather than practice, and follow *abstract teachings* rather than the *living example* of Christ, because Christ is considered as an *external* instead of a *spiritual Savior;* because men prefer to bow before the cross, the emblem of vicarious sufferings, rather than take *it up* and sacrifice selfish ease, and become a living, spiritual force in the earth.

Now, kind reader, you will notice, in reviewing these remarkable statements, that the answer to my first question was fully met, viz: "the setting forth of a perfect system of theology." The errors which the Christian world has fallen into seemed to me to be met in the statement that the commandment given under the patriarchal and Mosaic system, should not be confounded with the Christian system of truth, and the carrying out of Christ's laws of love contains the perfect divinity, the perfect system of truth. As Paul says, "Love is the fulfilling of the law; love worketh no ill to his neighbor." But my second question, with regard to the holy spirit power, was not fully answered, at least to my satisfaction. I considered that a system of truth, even though perfect, was not all that the world needed—but a moral power, a quickening, moral force, to cause men to feel and act at once under its all-searching scrutiny. Let the steam engine, with its simple combination of levers, wheels, and axles, represent the system of truth; then the fire, which generates the steam, sets all in motion, and gives force and power to all its parts, represents the moral force necessary to make the truth quick and powerful—even to the discernment of the thoughts and secret motives of the soul, forcing conviction and reformation on all whom it may reach. The Savior said to his disciples, "Tarry ye at Jerusalem until ye are endowed with power from on high." Here it is distinctly announced that beside the system of truth which he had taught them there was requisite to their success a power from on high—a Holy Ghost power. This power they were to wait for and receive before entering upon their mission of preaching the gospel system of truth to every creature.

Now, under a new dispensation, I considered that we might reasonably expect a greater power than accompanied the gospel at its advent, a moral power as much greater than that which ushered in or accompanied the gospel of Christ, as the gospel of Christ was superior in moral power to the Jewish or Mosaic, or the Mosaic superior to the Patriarchal dispensation. So I felt that distinct information on the all important question of moral power was of the most vital interest. To control and marshal moral forces around a system of truth, I conceived to be even of greater importance than to have a perfect system of truth, just as it is more important to regard the power that wields the sword, than to regard a flaw in the sword itself.

In the light of these considerations, I asked for direct or positive instruction on this, to my mind, the greatest and most important of all subjects connected with a new dispensation of grace to man. This was given me in these words from Christ himself, or one who personated him.

CHAPTER XI.

"The great mission to which you have been appointed does require that you should be in possession of a power sufficient to put in operation all the force necessary to *convert* the *world* to *God*—in other words, to 'subdue all things to *himself*.' The power which you desire is it not manifest, in part, in the fact of your ability to write upon every subject within the range of human thought without a moment's reflection? A power which you confess most potent and wonderful? Besides, already you possess the power of making others feel the *truth*; it is all *contained* within *your* own organization. You are now filled with the *Holy Ghost*—a divine power most potent, and all-sufficient to make any system of truth answer its ends.

"When you ask for a greater moral power than accompanied the gospel at its commencement, you ask what you do not need, for the amount of moral power needed is that amount necessary to accomplish the end, to-wit: the conversion of the *world to God.*

"Now if any man of pure nature, with the gifts of inspiration, and filled with the Holy Spirit, is not a sufficient moral power among men, then he ought to retire as did Elijah to the Cave of Horeb.

"You in the earth life do but faintly understand the immense moral power a devoted, pure, and good man is capable of exercising, especially when filled with the Holy Ghost sent *down from God,* (the Holy Ghost is identical with a Holy Spirit.) I say you can not, in your present position, fully understand the moral power a man in the form, filled with the Holy Spirit, can exert over the hearts and conscience of his fellow men.

"You now possess the true power of the Holy Ghost, and a hundred men like yourself, filled as you are with the holy spirit, would be a formidable force which no number could gainsay, or stand in *opposition* to. Moral forces are *gathered* by being brought in contact *with other moral forces;* one praying man is a moral force, because his desires go out into the thought atmosphere, and affect the minds and hearts of those who are prayed for. So then the moral power of a hundred holy men, with earnest purposes praying to God for blessings on the sinner, is increased not a hundred fold, but a hundred times one hundred fold, which makes the moral force absolutly irresistible.

"Is it not written, 'that one shall chase a thousand, and two put *ten thousand* to flight?' A revival power, as it is termed, is a *real power;* it is commenced by one or two uniting to pray. By so uniting a moral impetus is given. By increasing the number, greater power is manifested, and so on the principle is of universal application. If the whole of Christendom were united in earnest, persevering prayer for a special object, (such as the conversion of the world to God,) it would be done speedily. Look at the revival of 1857, when almost every Christian heart was an altar of prayer. What was the result. Thousands were converted to God, and made to rejoice in hope of the *glory* of *God*. In the day of the apostles, the same moral power was manifested. The divine power realized and felt on the day of Pentecost was accumulated by a ten day's prayer-meeting. When that earnest assembly came together, with hearts all aglow, and each filled with the Holy Spirit, they were a completely organized moral force—a moral battery from which was discharged a moral power, equal to the causing of thousands to cry out, saying, 'Men and brethren what must we do to be saved? And when told to repent and believe in the Lord Jesus Christ, as their Savior, they cried aloud 'We do believe,' and immediately they, too, were filled with the Holy Ghost.

"It is the law of the universe, that every human being in the form has a guardian spirit, or what ought to be a guardian spirit, but as often, when not holy, a destructive spirit. But all are surrounded by invisible beings, either good or bad, pure or impure, conscious or unconscious—it makes not the least difference, they are present. So you will understand distinctly that a revival power is only *the power to cast out Devils*, on a large scale. One good man may have the power to command an evil spirit to depart, and may by his presence, keep that spirit away; but remove the person from the presence of the good man, and he falls back to the same crimes that he formerly committed under the promptings of the evil spirit. You ask for definite information. You shall have it.

"The evil spirits are all around you. Their abiding place until the judgment, is the air, or atmosphere of the earth; and they work in the children of disobedience by impressing, or infusing into their minds, thoughts and desires which are wicked, and selfish, and also by inflaming their passions, thus leading them often to commit the most attrocious deeds. Holy Spirits are in paradise, which is above, and beyond the atmosphere of the earth. In order when they visit the earth, to give them power in any degree over the hearts and consciences of men, and to enable them to aid in fitting souls for heaven, it is necessary that there should be earnest prayer, real agonizing prayer, or the part of those in the form. The Kingdom of Heaven suffereth violence, and the violent, or the realy earnest, take it by force that is, by forcing the evil spirits to leave, and giving the ministering spirits power to aid in man's redemption.

"The injunction of one of the apostles, is 'to walk not after the flesh but after the spirits,' (which should have been after the Holy Spirit from God or of God.) When the man who has been brought, by the mighty power of God and the aid of ministering spirits, to yield submission in all things, then he realizes peace, and joy in the Holy Ghost, but when he relapses to wallow in the mire of sin, the ministering spirits are forced to leave him. This the Apostle calls quenching the spirit, whereby ye are sealed unto the day of redemption, or unto the day when ye shall depart from this life, as at that time, such individuals pass up through the region of evil spirits, which is the air, to the paradise of God beyond, there to realize that inexpressible joy, which is full of glory, and full of *real earnest* desire to visit the earth life, (when conditions are right) to aid in ministering to poor penitent sinners, who may be calling for aid against some of the demons, who may be dragging them down to the pit of woe.

"There is in this world a continual conflict, as you are aware, between the powers of darkness so-called, and the powers of light; between the regenerate and the unregenerate; and this conflict will continue till the end of time. You have asked, often asked, why the world has progressed so slow in godliness— why my gospel has exerted so little power. I answer, it is because you do not understand the immense resisting power of evil spirits. They are mighty, and it requires on the part of the good, and the pure, a constant effort to reach and help the poor captive children of men. It is only when a great revival movement takes place that any great number of ministering spirits can approach the earth life to exert their power in ministering to the wants of needy, *captive* sinners. Therefore, I am the more anxious that you should understand the law of *revival power*. If all the disciples would follow closely the directions here given, a vast influence for good might be exerted; an influence that would shake the kingdom of darkness, and soon bring it to an end.

"The slow progress made by my gospel in the world, since my ascension may be inferred from what has been said above—especially in the heathen world. There the powers of evil have complete sway, and it is not until many have been converted, and have passed away, that any great good can be accomplished. Every soul converted and received up into glory increases the power for good, because they return and in a measure form a wall of protection around their friends in the form, and though invisible to the evil spirits, yet they are nevertheless often present to soothe, cheer, and comfort by gentle impression.

"My beloved servant, Judson, spent many years in heathen lands before a single convert was rescued and brought to God. After which, another and another were added, until a little army of praying souls were gathered, and every praying soul is a living moral force. So their moral influence increased. And when others of the converts had left the church militant, for the church triumphant, the moral force was still farther augmented. As I said before, the kingdom of heaven was like a little leaven hid in three measures of meal, till the whole was leavened; each particle as it becomes leavened helping in turn to leaven that which lies adjoining it, and so on until all is leavened.

"You seem to be in doubt about *your calling;* you need not be *for a moment.* Shove off your little barque into the broad ocean of God's infinite love, and you will find aid and comfort. The true secret of success in any enterprise is to be fired with *zeal.* You need not fear to commence. You now understand the law of control in reference to moral forces, which may be brought to bear on the children of men.

"In conclusion, I will say that I design, first, to call in the Jews; then afterward the gentiles. So you will first proceed to New York. There you will be aided by H. W. Beecher and others, and then proceed to Palestine, where I will meet you and give you success."

While meditating on these things that "still small voice" came to me again saying, "I am the Lord your God, and have a work for you to perform. I must call in my people, the Jews, preparatory to that great event which is soon to transpire, and I have arranged for you to go to New York, and there meet H. W. Beecher, who will greatly aid you in your work of assembling, my ancient people at Jerusalem, preparatory to their being converted, and made meet for the kingdom of heaven. You will leave your home unknown to your family, and proceed to a place that I will direct you to."

CHAPTER XII

Upon receipt of this intelligence I said in my heart as did Paul, "I will confer not with flesh and blood," and immediately obeyed the heavenly vision or voice. So taking my best apparel, I passed to a buildidg in the rear of the main dwelling, where, after arranging for the journey, I proceeded to the place designated by the heavenly voice. On the way the voice continued to comfort me and gave me every assurance of protection. Passing down through the lower part of the city to the river, I was directed to go to a certain tree on the opposite side, which I reached by crossing the railroad bridge. There I was commanded to take off my shoes, which being done I was sealed with seven seals. Though they were not visible, I was assured that they would all appear in the fullness of time.

Having been sealed, I was then commanded to take the cars and proceed on my journey, which I did, arriving at the town of ———. I was directed to stop at the principal hotel, and there remain until further instructions.

At the hotel I called for a room and a good fire, and being seated, I fell into this train of reflection: Here am I, a weak mortal, called of God to leave home and friends and all that I hold most dear, to perform a great work. How can I sacrifice my home, my little ones, and all I hold so dear? But my mind soon settled the question in this way: God, my father, has called me; such an amount of evidence has been presented that no sane mind could possibly set aside, and if God calls, who of all the children of men can refuse? If God be for me, who can be against me. I shall prosper, though all the world may be arrayed in opposition.

I could not possibly persuade myself that any being, mortal or immortal, could be so vile as to impose upon me through and by means of the religious instincts of my being. I had somewhere read that it was possible for the devil to transform himself into an angel of light, and in that capacity deceive the children of men. But though he might do that, still I could not persuade myself to believe that the prince of evil would assume to be God, and also be able to understand the pith and essence of the gospel, and could express love and attachment to our Savior, Jesus Christ. I thought, too, that I had complied with the injunction of the apostle John to prove the spirits by the test of their fellowship for Jesus Christ. They seemed to me to have given full evidence of fellowship with the Redeemer and his kingdom. But what seemed to me of more weight than all else beside, was the mysterious voice whispering to my inmost soul, saying, "I am the Almighty, I am your Heavenly Father, and the Father of all living; I have called you, and you must obey."

While thus reviewing the past, and the evidences on which my hopes were resting, the unearthly and most mysterious voice came to me again, saying, "I am the Lord, your maker, therefore take your shoes from off your feet, for the place on which thou standest is holy ground." I obeyed, and with profound and reverential awe I prepared myself to attend minutely to whatever should be commanded.

The voice continued, "I am the Judge of all the earth; I, the Lord, have called you, to warn all men to flee from the wrath which is to come. The judgment day is approaching, when all men, small and great, shall stand before my throne. Therefore be dilligent, be faithful, and do as I command you, and great shall be your reward. You have left those you loved, many of whom are in the bonds of iniquity. Now, therefore, fix your mind upon those whose conversion you so much desire, and while you pray I will seal conviction upon their hearts, and they shall be converted for your sake, even while you are speaking."

With emotions, deep and overwhelming, I fixed my mind upon one for whom I had felt unusual religious interest. I prayed until sobs and tears choked my utterance, when that still, small, solemn, whisper came to me with distinctness again, saying, "Your prayer is answered, your friend is converted, and is now rejoicing with joy unspeakable and full of glory."

"My child," said the solemn voice again, "fix your mind upon another, and I will bless again even for your sake." I commenced then again with the same childlike simplicity of prayer as before, fixing my mind distinctly on the person I desired converted, and after a few moment's of earnest pleading, I heard again the voice saying, "Child, thy prayer is heard and thy friend is now happy in my love."

Thus I continued for many hours, my heart becoming more and more interested in the work and swelling with the love of God, as I continued to fix my mind on one after another of those whose soul's salvation was near and dear to me. At length the solemn voice said, "Have you no other relations and friends on whom you can fix your mind in prayer?" I replied by saying, "I can think of no more." "Then," said the voice, "fix your mind upon any ungodly man you choose, and I will hear and bless him for your sake."

With tearful eyes I then renewed my prayer, fixing my mind upon one and then another of those whom I knew to be the worst in wickedness. At the end of each petition the answer came: "Your petition is heard, and he, for whom you supplicated is among the redeemed." This continued perhaps an hour longer, when growing weary from exhaustion and fatigue, I retired and slept as sweetly as a child upon the bosom of its mother.

As the soft light of morn stole into my window, I heard again the still, small voice, saying, "arise my child, and hear the good news. Those for whom you prayed are happy in my love, are rejoicing in hope, and have heard of your mission, and are coming to rejoice with you, and bid you God speed."

Having arranged my toilet, and kindled the fire, I ordered breakfast to be served in my room. When seated at the table, and about to commence my repast, the voice said in the kindest and sweetest manner: "You need feel no embarrassment in my presence; have I not always been present with you, do I not know your every thought, your every word, and your every deed? Ask the blessing on your repast in your usual way, just as if I were not personally present, for though I am not present everywhere in person, yet I am present everywhere in my omnipotent power and wisdom."

The repast being over, the voice of the Almighty, as I believed, said, "This day will be a day to this place such as they never experienced before. God is in this place, and they know it not. I will now whisper to every man's mind, as I whispered to you in the still small voice, saying, 'prepare to meet thy God in judgment,' and hearing this they will assemble in the different churches for prayer; at the same time the converted ones for whom you prayed during the night will have arrived in the place, and will join in the general rejoicing, and weeping, and crying for mercy, and thus the wave of salvation, so astonishingly begun in this place, shall roll over the whole earth, because that day is approaching, that great and terrible day, when all men shall receive according to the deeds done in the body. In the mean time you may go up and call on Mr. A——, who is my servant, and a godly man. You will find him at his church preparatory to the great assembling of the people."

I was so moved by this good news, and the promise of salvation to such multitudes of perishing souls, that tears flowed freely from my eyes. In this condition of mind I passed down the main entrance of the hotel to the street. Looking around I saw no unusual stir, but thinking that God worked silently with every heart, I passed on with the certain expectation that I should find the minister at the church designated, and many assembled for worship.

On arriving at the church I found, to my astonishment, the doors closed, and not a single person in or about the building. I soon found the minister at his residence, and to my still greater astonishment he informed me that there was to be no meeting there that day. I returned to the hotel, expecting that by this time those for whom I prayed and whom I believed were rejoicing in a conscious hope of sins forgiven, had arrived. But in that, again, was sorely disappointed.

Passing up to my room, I inquired of the

Lord, why this strange failure. To which the voice replied in the same distinct, and well defined whisper: "The failure is caused by the mischievous conduct of wicked spirits, who have, of late, been whispering in the ears of the people, which has confused them, and they do not recognize in my whisper the still voice of the Almighty."

"But," said I, "what will become of thy great name?"

The voice replied by saying, "I will take all remembrance of this failure from their minds; and they shall know that I am the Lord; that with me there is no variableness, or shadow of turning. But you, my child, will proceed on your mission of calling in the Jews, the same as though this seeming failure had not occurred."

"But," said I, in reply, "Father, I have not the wherewith to convey me to New York."

"Yes, my child, but I have provided against all contingencies of that kind, by impressing a wealthy man in the city of New York to telegraph the bank in this place to furnish you all necessary funds. Be therefore not faithless but believe."

This calmed my mind, reassured my confidence, and I immediately left for the bank. Stepping up to the counter, I inquired if a certain man, calling him by name, now living in New York, had telegraphed to this bank to place an amount of money to my credit. The banker assured me that no telegram had been received.

Again I inquired the cause of the failure. The same mysterious voice replied by saying, "The cause of this failure is the same that produced the others; but," continued the voice, "I am the Almighty. I have power to kill and make alive, and those who have interfered with my purposes I will judge; therefore rest in hope, and all shall be made right."

I yet had confidence in my senses. I was certain that I had heard the voice, and I could not force myself to believe that any creature above or beneath could be found who would dare to personify or assume to be the Almighty himself. I also thought back over the communications I had received. I reminded myself of the deep toned piety which pervaded them, and of the kind assurances given me; and summing it all up, I felt doubly impressed that I could not be deceived.

While thus meditating, the voice uttered these words, "Return to your home, and all will be well." Obedient to the command, I immediately set out for the place of my former residence.

My sudden disappearance from home had caused no small stir among the friends and relatives, but my presence soon reassured them. How little, thought I, do they understand the real cause of my absence.

The reader might suppose that the would-be divine intelligences who had followed me so long and had so grossly deceived me would, upon having been discovered to be but devils clothed as angels of light, have left me never to appear again. But it was not the case. So far as ability to impress my mind with their thought was concerned, I found that they possessed even more power, and that it was every day increasing.

As soon as it was really apparent that I had been deceived, I sunk into the very grave of disappointment. My hopes, which had been raised up to the seventh heaven, were dashed down to the lowest pit of hell. My invisible deceivers for several days continued to flatter me at times that all was well—that God was as really in the darkness of this disappointment as in the light of the brightest hopes of former days.

Said one of them to me, in an assuring manner, "God has permitted all this for the trial of your faith—to test your steadfastness under adverse winds as well as under favoring gales. You ought to remember that Jesus, your brother, before he entered upon his glorious mission, was led up of the spirit into the wilderness, to be tempted by the devil. In other words, he was turned over to the rulers of evil that his mediumistic powers might be developed—that is, his mind rendered more susceptible and impressible, so that angels and the highest order of intelligent spirits might approach and communicate with him."

This, they assured me, would be the case with myself; that as soon as I had been sufficiently tried and proved to be true under the most adverse circumstances, then I would be allowed to be brought in contact with the holy spirits again. They also assured me that though they were wicked, still they were engaged in the very holy work of developing my mediumistic powers; that they were appointed to that work from the fact that their magnetism was stronger than that of the holy spirits; that they could do as much in one day toward the development of my mediumistic powers as a holy spirit could do in many months. They further assured me that holy spirits superintended the whole work. They reminded me that often in the earth life men, in rearing the loftiest and most sublime aritectural structures, find it convenient and profitable to employ men of the very lowest order of mind to do the excavation for the foundation, and also in preparing and laying that foundation. So in my own case I had been turned over to them, only that a foundation might be laid upon which to rear the loftiest superstructure of mediumistic power. "Have faith," they said, "and all will be well."

Thus at times my hopes were raised to the highest pinnacle, only to be again dashed to the earth.

CHAPTER XII.

About this time I commenced to feel a pain in my temples, at times extending down to my cheek bones. This would continue for hours and then cease, then after the lapse of a few hours it would again begin. Sometimes the pain was almost equal to that caused by the extraction of a tooth, then at other times it would be quite moderate. I inquired what could be the cause of these pains, whether they proceeded from neuralgia or some other nervous disease. They replied by saying that I would find out by-and-by, only have faith as a grain of mustard seed. But this reply did not satisfy me; I repeatedly importuned them to inform me.

I had a vague impression that they were developing my powers as a seeing medium. After a time they frankly informed me that I would make a splendid seeing medium, and to reward me for all my suffering I should see all my deceased relatives at once and converse with them face to face. This I thought would be an exceedingly nice affair. I therefore concluded to patiently bear the pain, severe as it was at times, besides I had no power to stop their proceedings, and therefore I the more readily allowed them to proceed. But at times the pain was so severe that I shrunk from it, and tried repeatedly to prevent their operations. This I attempted to do by putting my head under a pillow, knowing that feathers were a non-conductor of electricity. This not succeeding, I tried the experiment of placing a large deep glass dish upon my head, but all to no purpose. The work went on, and as I could not prevent their operations, I resigned myself to their tender mercies. I found, however, that expostulation succeeded better than trying to place physical impediments in their way, and occasionally at my earnest request they would desist, and give me a season of perfect quiet.

At times I felt anxious to know who these spirit friends, as they still called themselves, could be. I therefore requested that each should give me a short biographical sketch of the life they lived in this world, and also the one they were now inhabiting. With this request they seemed perfectly willing to comply. I therefore procured some paper and they dictated the following as a part of their history:

"I made my exit from the body and came to this world some twelve years ago. Was, before I left the world, a member of an orthodox church, and greatly beloved by all the membership, and especially by the class of fine lads I then taught in the Sabbath-school. Elder W——, my pastor, said, in regard to my standing in the church and community, that I was one of those rare young men that anybody could approach and not feel the sting of harshness; that he loved me as a brother, and would give all that he possessed to know me as his own natural brother.

"After arriving in this world and finding that I had missed heaven and was doomed to remain with all the wicked of earth life, being reserved under spiritual darkness unto the judgment, I therefore resolved to play the devil generally. So I commenced roaming around among the dens of infamy to which we have access as well in this world as in the earth life. These dens of infamy and vice I often entered, and though at first there was much to shock my modesty and moral feelings, yet I found it true here as in the earth life that vice is a monster of such hideous mien that to be dreaded it must be seen, but seen too oft familiar with her face, is first pited, then endured, then embraced.

"In these places of wicked resort I met many of those with whom I was acquainted in my earth life. These I found were in the same condition with myself—had missed heaven, and as all was lost they, as well as myself, concluded to do up a large amount of the devil's work in as short a period of time as possible.

"Finding you in a condition to impress with considerable ease, we commenced the work of leading you to believe that all the modern spirit communications were of God and all superintended by angels, and would in the end bless the whole earth by its wonderful manifestations. Thus you were prepared to consider that whatever emanated from that source should be interpreted in some way as working together for the great and ultimate good of all.

"Your mind was gradually prepared by mesmeric influences to receive more perfectly our impressions. We continued this course of treatment until we had gained such crontol over you as to slightly influence your hand to move involuntarily. In the course of time we were enabled to insinuate into your mind a full idea, and so on until we became master, to some extent, of your whole being. We then made arrangements with a number of what we call wise spirits (though you would not call them wise, because not good,) to write the first communications which were received at the commencement of your course as a spirit medium. They contain the very essence of the gospel of Christ. But from this learn that the devil can preach the gospel, and eloquently too, when he has an end in view. This we found succeeded so well in gaining your confidence that we concluded to go farther. Of one thing we were in doubt, which was that you were a Christian at all; we therefore determined to prove you. We were also in doubt whether there was any true piety in the world. We had doubts of this from the fact that in our own case we thought that we possessed true piety. We prayed, but more from a sense of duty than otherwise. We attended church, but more from habit than from a desire to glorify God in the salvation of sinners. We taught a class in Sabbath-school, but it was more because asked to do so by the superintendent than from a desire to do the children good.

I was polite and affable, because it gained for me the esteem of all with whom I associated. I kept out of bad company, not because I disliked it, but because I feared the consequences;—loss of reputation, loss of the esteem of others, and I had no higher motive in doing the things which I did do than the motive ot gain in some way. God's glory, the good of the race, the final salvation of a soul, never enteren my mind. I found that all my hopes of salvation were therefore cut short, and as if to spite the Almighty for not sending me an effectual call, such as would render my salvation certain, we have done in part what we have done to you.

"We commenced, as you are well aware, by making you think that God had called you to become the second Christ.

We had our fun in seeing you really and sincerely pray for sinners. We were delighted, very greatly delighted, to witness your extreme gratification at the conversion of your friends, and neighbors. We now know that those were genuine, hearty tears. We know that any man who will stay awake, and pray with tears streaming down his cheeks, nearly a whole night, for the salvation of friends and relatives, is certainly of the true, genuine material, of which all Christians should be made.

"Then we were pleased by your evincing so much faith in God. Your faith has as far exceeded faithful Abraham's, as his faith exceeded that of Peter. Because first, you left a home of the highest type of excellence, with wife, beautiful and happy children, and an endeared circle of friends. You left, too, without money, and without scrip, at the bidding of one whom you had reason to believe was the Lord, your maker.

"Second, you prayed for those who were your bitter enemies, and in the simplicity of your faith believed them converted.

"Third, You adhered to God, after he had deceived you, and given the evidence of moral weakness."

What to think of this communication, was more than I could determine. That it was true in reference to the circumstances referred o in regard to myself, I felt absolutely certain, but whether the reasons assigned for treating me thus were true, it was impossible for me, under the circumstances, to determine. The story on the face of it seemed very natural. If it were all true, I could but say that God, my Heavenly Father, had tried my integrity, as he, of old, tried and proved that of his servant Job, only instead of three adversaries, there appeared to be legions of them.

Having obtained an answer as I thought to my first question, I continued to venture and ask another. I then said, "Can you inform me whether there are real devils and a Hell, and if so where are they located? To which I obtained the following answer:

"We say, in answer to the first question, that we have not seen any devils since we arrived in this world from your earth. That there may be devils unseen by us is very likely, and for this reason: We are invisible to you and yet you have evidence that we exist. So by the same method of reasoning there may be devils and we not see them. It would seem, judging from the way we have treated you, that we have been posessed of a legion of them. If they have the power to impress us and speak to us as we do to you, we have no knowledge of their having done it. Still they may have the power of infusing their subtilty into our minds without actually speaking to us—just as we can infuse our thoughts of evil into your mind without uttering them in distinct words. You are well aware that before you became so impressible as to allow us to speak to you that you realized ideas coming into your mind even when uncalled for. So we often realize the same thing, but whether they come from devils or not we can no more tell than you can tell that we *are or are not devils.*

"You ask us if there is a hell, and if so, where it is located. We say that there is a hell more awful than the human mind in the form can conceive of; a hell not of fire, though inconceivably worse than fire to us; a hell that receives all the moral filth of the earth life; a hell more terrible than fire, because fire can not burn spirit in the least or cause to a spirit the least possibly amount of suffering. The suffering that we, and all others who abuse their privileges, are in is greater than we can bear. We pray for death as you have heard us often—not temporal death, but eternal extinction of being. The greatest boon we crave, is actual annihilation. The parable of the rich man and Lazarus is not anything like a clear and vivid representation of the reality. That there is an impassable gulf between us and heaven is absolutly certain—a gulf so wide and deep that none ever passed it. I am told that heaven is somewhere out side the earth's atmosphere and beyond the orbit of the moon. If so, then you understand where the gulf is that I allude to. We are confined to the earth's atmosphere by a law as fixed as the thorne of the Eternal—a law as operative as the law gravity upon your bodies in the earth life.

"You will know, then, that the air, the earth's atmosphere, is the abode of the wicked, and is the only hell we know of. There may be as many hells as there are worlds in the Universe for ought we know. This hell, with its myriads of inhabitants, may be the abode of the wicked only until the great day, judgment—then its locality may be removed and the conditions of it changed. But so long as there is to be found a man on earth whose heart is fully set in him to do evil there will remain this hell to receive him. The scriptures give as clear a representation of hell as it is possible for human language to convey to the human mind. The only false view that it apparently gives is the representations that the sufferings of hell result from being cast into a lake of fire and brimstone, which is not the case, at least not immediately after death. That this world will perish by fire is not only the assurance of Holy Writ, but the legitimate deduction of reason, and our earnest desire is that we may perish with it.

"It may seem strange to you that the Almighty should have created a hell,—but it is not at all strange to us. The inhabitants of hell are the legitimate fruits of wicked and corrupt earth life. A man must reap the rewards of his own hands by a law as unchangable as the Almighty. Every man must be judged according to the deeds done in the body, which we now know by experience means that man is and must be essentially in the spirit life what he was in the earth life. Man, by nature, is a demon, and grace alone makes him a saint. Therefore, unless a man realizes that change which is from death unto life whilst he yet remains upon the earth there is and can be no realization of it here, among the wicked which surround the earth, who constitute the inhabitant and powers of the air.

"The laws of eternal life are laid down in the gospel by the Lord Jesus Christ. He declares that the man who believes on me (that is on Christ) hath everlasting or eternal life and he will raise him up, (that is, his body), at the last day. This we now know means that he will give them eternal life, because they believe or because they have realized as in your case that inward evidence of it, which alone the Holy Spirit gives. No man can really believe without evidence, therefore no man can believe in Jesus Christ but he to whom the Father reveals him; I mean in his moral excellence. Therefore, all men who do not attain to eternal life through his son,

Jesus Christ, will and of course must perish.

"That this earth is to be dissolved and burned up may be inferred from the fact that stars have already been known to go out, and reasoning from analogy this earth is as likely to dissolve with fervent heat whenever the breath of the Almighty is blown upon it, as the stars and planets already seem to perish. The great mistake of the bible reader consist in their literalizing in reference to the future of both the conditions of the righteous and the wicked. Whereas, figures of speech were used to represent as near as possible spiritual conditions, every one must in a measure understand the difficulty of describing spiritual states and conditions by comparing them to states and conditions on the earth. The expression, banished from the presence of God and the glory of his power' is very expressive if you in the earth life, could in the least degree understand what is meant by it.

"The human mind, from its very constitution, must localize and literalize. Jesus taught that the Kingdom of Heaven is within you and depends upon the disposition and tempers of the soul. This is pre-eminently so. The intense sufferings in this world consist in giving vent to unrestrained tempers, to exhibitions of pride, and the most intense selfishness. We understand now, as we never could know in the earth life, that Heaven as well as Hell is a place—though conditions of mind and heart are very much, yet *not all*. A man whose heart is right in the sight of God enjoys peace within, but as your experience will testify it is not always so without you are surrounded by those who could drag all heaven down into the mire and filth of hell to gratify their desires, whims, and caprices. Neither you, nor any one in the earth life, can have any thing like an adequate conception of a condition of society where every abomination and wickness is practiced; where supreme selfishness is the universal rule of action; where all are influenced by one motive and that motive is self, and self only.

"Your own experience can fully testify to this fact, that the inhabitants of this world can actually delight in the torture of any one or any thing without showing the least mercy or forbearance.

"There is such a thing here as in the earth life of *fun* wearing out. In your case we not only wore out the fun, but your system became so innured to electrical discharges that we lost the power to control or influence you. The nerves of your system became, by continued discharges of electricity upon them, like copper wires capable of conducting off the largest discharges of electricity, with but little effect to your physical system. Thus you may know that the lion of evil is chained."

CHAPTER XIII.

This communication astonished me more than the other. Because, first, though the communicator did not acknowledge that he was a devil, yet he seems to reason from analogy that there are such beings who infuse a deadly venom into their minds. This account agrees with the scripture account of devils and demons, and is as true as human language can express it.

My opinion of devils and demons was made up from reading the scriptures. I think they teach that fallen angels are the only actual devils, and that the term demon is the word which ought to be used when speaking of wicked persons who have left the earth life.

My mind became very much interested to know whether these beings around me were actually devils or demons. I, therefore, resolved to notice particularly their every word. As I could distinctly hear them talk to one another, they seemed to have thrown off all disguise, and very properly, I think, concluded that they must reveal sooner or later their true characters. The spirit's description of hell, I think, may be said to be graphic enough to have been written by one who had realized the whole length and breadth, highth and depth of that dread abode. The location of hell did not accord with my ideas at all. I had supposed that it was located in the regions of space beyond the outskirts of creation, in what might be called chaos, and that the few wicked spirits that surrounded earth had made their escape from there, and of course were moving around in the air and doing all the mischief they could find to do. The prince of the power of the air working in the children of disobedience, I did not understand as entering into and taking full possession and control of their bodies and minds, but infusing into their minds, by impression, their evil designs. That they did even take possession of some in our Savior's day, and control them just as a mesmerized person is controlled by the mesmerizer, I have not the least doubt; but I had formed the opinion that when Christ, by his gospel and spiritual presence, had taken possession of the whole earth, that then the demons and devils would all be driven into *outer darkness*, and there remain forever, and that this constituted the prison house of God's universe.

The reader will, of course, understand that any opinion advanced by the intelligence who dictated this communication is simply his opinions, and not entitled to any credit only so far as they agree with and throw light upon the bible. From expressions dropped in the two last communications I think we may conclude that lost souls in hell have no means of knowing their future destiny, only as the bible reveals it. That they desire and pray for annihilation, I know is true, but that they have any means of ascertaining their final end, except from the bible, is also true. That there is every variety of opinions held among these legions of the wicked one upon the subject of their present and future destiny I am fully persuaded from more than four years' experience with them. But the fact of obtaining eternal life through believing in Jesus Christ; that no man can believe without evidence is virtually true. From the deductions of reason, and the declarations of scripture, I positively know from experience that there is a revelation of Jesus Christ in the soul. I know that once Jesus Christ was to me as a root out of dry ground, without moral beauty or moral excellence. But now I know that he is the highest conception of what I desire to be. I can truly say, that as "The heart panteth after the living waterbrooks, so panteth my soul after thee, O God."

This spirit seems now to understand the way of life through Jesus Christ. Whether he obtained it by torturing me to the very verge of the grave, I know not. He says it was done for that purpose, in the first communication. But my own opinion is that he was one who laid this plan, deceived and tortured me for his own amusement, for the mere gratification it afforded him.

Repeatedly I charged them with being devils, because they could take pure delight in causing the most extreme sufferings to one who had never injured them in the least. To this they replied by saying they were devils, but did not like to be called devils. I could not persuade myself that they had ever been in the earth life, for the reason that we have nothing with us that corresponds to the degree of wickedness which they exhibited. A savage of the wilderness will torture an enemy, but was never known to torture one who was friendly or had not in the least injured him. It is almost impossible for the human mind to concieve of the extreme wickedness which would lead one being to torture another without having given in some way offense.

I reasoned that if these beings had ever been in the earth life, they had upon reaching that world grown suddenly more wicked. As a means of increasing my mental torture, some of them claimed to be my relatives; others that they had been my friends and neighbors, but had failed in reaching heaven, and now, because they could, and as if to spite the Almighty, they had determined to torture me to the extreme of their ability.

I was certain that if these wicked spirits could talk to me and dictate their desires, opinions and wishes, that holy spirits could also do the same. I felt hopeful, therefore, that in time they would come and cause these wicked spirits to leave me. Besides, I had made special supplication to God, to the end that he would deliver me from these wicked spirits, and give me peace and quiet, as in former days. This prayer I certainly expected God would answer, and therefore looked for peace. But, dear reader, no peace came. No sooner did my weary head strike the pillow than the electric shocks commenced and continued for a time, and then ceased entirely. All was silent, not a whisper was heard. With this I was pleased, I began to lift up my heart to God in prayer; with a grateful sense of his goodness that my prayer was answered, and my soul relieved from its demoniacal surroundings. O, thought I, what a comfort to be alone; to have my own thoughts, to be free from sharp, unfriendly criticisms at every turn the mind might take. Oh, I exultantly said, how sweet and delightful it is to be alone again.

But the lull, the sweet calm, was only the precursor of the more terrible storm. Soon I began to discern about the room flashes of light; then I caught sight of cloudy forms moving to and fro; then I discovered that some objects in the room which were distinguishable by moonlight, wore a hazy glow of light whilst other objects looked as they usually did. Then, sudden as the lightning's flash the whole room appeared lit up, and arranged in rows around the room a thousand human skulls, tier above tier. If they were attached to skeleton bodies the bodies did not appear, as the front rank began so low down that I could not see them distinctly. After continuing a few moments, well, thought I, (the shock of the moment had subsided,) that certainly is a strange sight—is it real or illusory.

I then gazed around the room to assure myself that I was not asleep, nor dreaming. I looked at the different objects in the room and all remained as they were. I then knew that it was a *real, positive, actual* sight.

What, thought I, do wicked spirits assume the form of skeletons in the other world? It can not be. At that moment another flash, and the whole room presented the same appearance again. Oh, horrible sight! a thousand skeleton forms, gazing from hollow eyes. Silence reigned! I looked, gazed, and then said to myself, this is real, and as I seemed to pronounce the word, it was all dashed out, and the room remained as before.

I then began to muse upon the probable form or shape that wicked spirits assume after making their exit from the body. I had received the impression from some source that wicked spirits, when emerging from the body, assume the human form, only that they were a more or less horrible aspect as they were more or less wicked.

In this manner I was musing when another flash of light revealed a vast, terrific, *massive, black looking* form, with fierce flashing eyes, and a monster mouth. It stood as it were looking right down into my very soul. I turned as if to get a better view, to assure myself that it was real, when the familiar, whispering voice, said, "His Majesty, the Devil." This monster object stared at me for several minutes, smacking his lips and showing his teeth, then disappeared as suddenly as the visions of skeletons.

I wondered how such horrible objects could be produced. That they were real, my reason decided could not be; if they were phantom images, I said, by what power are they produced? That they were produced by spirits, with a design to alarm and startle me I was fully confident, but how such real, life-like pictures could be created so instantly, was to me a deep mystery. At first sight, I was a little startled, but as soon as I thought a moment I felt calm. I reasoned thus: God is the author of all things; if he has created creatures he can control them, and if he sees proper to take me out of the world by their instrumentality I was willing to go. I felt a sweet assurance that he loved me, and if it was his will that I should be destroyed by these wicked, invisible spirits, then I should submit as calmly as by any other means which might seem to him proper.

Thus a few moments passed on and again the flash, and a huge blacksmith appeared, with brawny arms and hammer uplifted, forming a perfectly life-like picture. It remained for a few moments, and while viewing his huge proportions the voice came in distinct and measured tones, "Vulcan, the Blacksmith," and then he disappeared.

Thus one scene after another came to view in the usual order of panoramic processions. First, a ludicrous scene with strange, comic, figures, more grotesque and absurd than is possible for the human mind to conceive of; then a change would bring to view a motley crowd—some black, some white, some sober, and some grinning. Thus the views passed on for hours, sometimes reptiles the most loathsome and of the strangest form and colors altogether different from any thing described in natural history. At one time a huge boar, with monstrous tusks, came dashing right up almost into my face, and there stopped looking me fiercely in the eye. I reached to drive it away but I found it remained. I looked at the window and then at other objects in the room, and then at the picture. I found it so real, positive, and lifelike that I shrunk back from waging a warfare with such a terrible creature. A lion, then a tiger, then other animals came dashing in turn almost into my face, fierce as ten furies.

So they continued until I became exceedingly weary, and begged of them to cease their operations, but all to no purpose. Occasionally I could hear them say "soon we will close. These are the animals and creatures found in spirit land. We want you to see them." Thus after many hours they retired and all was quiet. I then obtained a little rest.

Thus, dear reader, you have a very faint and imperfect description of a spirit panoramma. That you may be assured that this is no fiction of the brain, I would say that it has continued with short intervals until the present time (a period of over four years.) These scenes have presented a great variety of changes — sometimes they produced the inner court of some gorgeous temple; then trellis work hung with the most luxuriant vintage; then perhaps it would change to an exhibition of the finest lace tapestries, the most costly robes and wearing apparel, also the finest stucco, medalion, and fresco work which appeared in bold relief upon the walls. These latter were produced generally in the morning in full daylight.

I sought to know how they were created and this was the answer they returned :

CHAPTER X.

"These mind-pictures are produced psychologically. The image is not produced in the room but in the mind—produced there by the will of the operator. We at first designed to frighten and terrify the subject, but finding that his nerves were made of sterner material, we changed our tactics to that of something more pleasing. We designed at first to introduce his relatives as among the psychological characters, but finding that he was aware how they were created —being not real but mental pictures—we therefore partially dropped the work of psychological manifestations and commenced the work of what is called general developments. This is accomplished by passing currents of electricity along the leading and sometimes the minor nerves. These nerves are to the body what the wires are to the electric battery. The principal object of this so-called development is to give the nerves capacity to conduct electricity without apparent shock to the physical system. As in your case, development is often attended with great pain. It is often performed, too, by those who very imperfectly understand the work, and then again by those who are rough and have little regard for the suffering of the subject. During the development of this subject we have inflicted much physical suffering, but not more than is often experienced by persons during a course of severe illness.

"We have no claims upon him for the patience he has manifested. That he has the grace of patience to a marvelous degree is shown in the fact that his life came very near being taken several times while making some of the experiments, especially when we produced the characters of one of the heavenly host in full dress. We might say, in partial justification of the course pursued, that we expected at first to have rendered the subject a splendid trance medium, but failing, we became reckless of everything like decency or justice, and abused him without stint or mercy just because we could. I would say further that unless the whole trickery of spirit seeing is exposed the world will be wretched indeed. The whole matter of spirit seeing in the modern times is an imposition upon the credulity of the public. The visions of A. J. Davis, as well as those of Emanuel Swedenberg, are bare-faced deceptions, so bare and transparent a falsehood that the world ought to see through the cheat at a glance. This subject, as well as many others, has suffered vastly more than many do in the soul's separation from the body. He has suffered mentally sufficient to have caused in many insanity, but without the least apparent derangement of the mental faculties. We attribute his patience and mental stability to simple trust in God. We know of hundreds of cases where persons have been driven hopelessly insane both in America and Europe by mesmerization of spirits, many of whom are as ignorant of its laws as the nursing child. Not always wilfully doing the mischief but nevertheless doing it. This, with many other abuses, has rendered the subject of spiritualism most odious in the minds of all who are not deceived and carried away by its glittering fascinations."

Thus, reader, you have the confession of what I believe to have been one of the spirits who deceived me. Whether these pictures were, as he says, mental pictures I know not; but to me at first they were as real as life itself. After a short time had elapsed, I had a conviction that they were not real, but mere mesmeric impressions, or something of that kind. I repeatedly entreated them, by every possible argument that I could use, to desist from making their impressions upon my mind, and from causing in different parts of my body such sharp, electric pains, but to no purpose.

Sometimes I would use this language: "Will you tell me why you persist in giving me so much pain?" This would be about the answer: "We give you pain because we can't develop your powers without it." Then I would say, "Why develop them at all?" The answer would be, "Because we desire to make some experiments in psychology." Then in reply I would say, "But what benefit will that be to you or to me, or to the world?" Then about this answer would be returned: "We wish the world to know that we exist, and that this subject is psychological." Then in reply I would say, "Is that a sufficient reason for inflicting on me so much pain? The fact that you can move my hand mechanically, and can dictate your ideas or thoughts to my mind is sufficient evidence of your existence." Then the reply would be given, "Very true, there seems to you but little use in psychological pictures, but to me it seems of considerable importance."

"First, it is the means which is resorted to all over the world to induce people to believe that they see spirit forms, when in fact they see only the mental pictures which the spirits are disposed to give them. Therefore, by this you will understand the fallacy of seeing spirits. Also, the imposition of clairvoyant seeing. You know by experience that spirit-seeing is an imposition—a mere trick, played off by impressing the subject with the scene which the spirit wishes him to have a view of, and then informing him that it is an actual sight of the place or person presented to view. For example: a sick room in London is, so to speak, photographed in the mind of the subject, who is then informed that he actually sees the person and surroundings in London, Or to make the matter still plainer, you remember that the Congress of the United States was seen by you in full assembly, and different characters engaged in earnest debate. You surely now know that this was only a photographic picture upon the mind. No man can see the actual essence of a spirit any more than he can see God. You are aware that your optic nerves are many times more sensitive now than formerly, and still with all this increased sensitiveness, you have not been able to see even the semblance of a spirit. You have seen cloudy appearances, but they were only the mist and vapor in the air attracted to the spirit, and thus giving it a hazy, misty appearance. At other times you have seen actual forms clothed in white, both in an out of church. Those were only appearances produced in the same way. The medallion and stucco work, though apparently fixed upon the ceiling of your room, is of the same character. You have suffered to be sure, but still, even if you have suffered greatly, it is, perhaps, a *little* consolation for you to know that many others have suffered as well as yourself. There is now known to be in different parts of Europe and America many scores of psycological subjects, but none so good as yourself. But is it necessary that one should suffer so intensely to simply make an exposure? If even one person had been made good or rendered morally better by it there would at least then be the semblance of a reason for the suffering, but that to my knowledge is not the case.

"But you forget that the world is running mad after the beast. [Rev. XIII: ii.] It has come suddenly upon the world, and they wonder with very great amazement and your timely warning, and very full experience, will save a vast number from the awful gulf into which they may be ready to plunge. Besides it has afforded us a vast amount of fun. Your case has been a very peculiar one. You were first led into a belief that spiritualism was but the harbinger of the millennial glory, by the few first communications. They were certainly very grand, and were given with the express design of leading you to believe they were from Jesus Christ, and God himself. You ought to have suspected this. All hooks are baited with a very gilded bait. But your honest heart could not but trust that all are genuine lovers of Jesus Christ, who are ready to say, 'Lord, Lord!' Now you know by experience that all is not gold that glitters. Besides, you also have a greater degree of patience than you had before—even patience under severe and long continued suffering. The abuses that you have so patiently borne were undeserved, unprovoked, and therefore there can be no excuse framed as a palliation of the offense."

CHAPTER XVI.

The reader will notice that the writer of this communication alludes to my seeing forms of angelic shape while sitting in church. I will say in reference to the matter, that I have been troubled very much with these psychological images, as they call them. During the time of prayers, while my eyes are usually covered to exclude exterior objects, the psychological light begins to appear, and to increase until it often exceeds in brightness the light of the sun. In the center of this luminous disk or circle a human form appears of exquisite mold, resembling statuary covered with thin silk gauze. To prevent this I resorted to the expedient of keeping my eyes open during the exercise of prayer. But at times the spectrum or vision would appear even when my eyes were wide open. Sometimes these images have assumed lewd and disgusting attitudes, very greatly distracting my mind from worship. The pain that accompanied these demonstrations was often very severe. The feeling was like a finger placed upon either side of the head and pressed with violence. This pain would sometimes change instantly from my temples to my cheek bones, then to the back of my head, then to the top of my head. This was continued until quite a preceptible soreness was felt in those regions. Sometimes an interval of two or three days elapsed between the painful operations. All my persuasions to desist were treated with harsh rebuffs, and as if to demonstrate the fact of their total depravity, they often endeavored to excite in my mind fears with regard to the safety of my life. Besides attempting to excite groundless fears, they seemed to tax their ingenuity to invent sharp, severe and cutting remarks. But that which proved the source of the greatest annoyance was their incessant talking, either to myself or to one another.

At one time I propounded to them this question: "Are you within or above me?" The answer returned was this: "We are sometimes absorbed into your being; at other times we are above and around you as best suits our purpose. Not more than two of us have ever been *en rapport* with you at the same instant. We mean by rapport—absorbed into your person just as water may be absorbed by a sponge."

This astonished me. It seemed at times to be true, as their thoughts seemed to flow into my mind as one drop of water flows into and mingles with another.

The reader may be anxious to know the subjects upon which they talked, and what replies were made. The following is a verbatim conversation:

QUESTION.—"Are you not in the truest sense demons? You seem to have no love towards Jesus Christ, and seem only determined to gratify self even at the expense of intense suffering to others."

ANSWER.—"You ask us whether we are not demons, and assign as a reason that we have no love for Jesus Christ, or any other good being. In this you seem to forget that in the earth life you have very many who had no love for any thing good, and are bent on gratifying self and only self, even at the expense of others. Surely, you can not expect this world where we are to have any thing better than proceeds from your own."

QUESTION.—"But we have no such characters in the earth life, either among barbarous or savage nations, characters who inflict pain simply to see the victim writhe in agony?"

ANSWER.—"Neither have we. You suppose that because we caused you such intense suffering that we had no motive but to see you writhe with pain. We had a motive, though we had no moral right. Our motive was fun, and fun indeed we had; and to us glorious fun, and much of it."

QUESTION.—"To say that you took intense pleasure in my sufferings, and had no other motive but the delight it afforded, you virtually confess the charge. Such a disposition of itself constitutes any being a devil.

ANSWER.—"But you forget that children often pull the wings and legs from the bodies of flies to see them writhe in agony, and therefore you would say we were using very harsh language, to say the least, to call them devils because they may do so."

QUESTION.—"Very true, there are those who inflict pain upon animals needlessly, and even take delight in seeing them writhe in agony. But we say that *that* is evidence of great cruelty. But an instance can not be cited where a parent ever so tortured a child or a friend, or even an unoffending stranger?"

ANSWER.—"Yes, there are many cases on record of parents so vile that they have even burned their children by slow fires, and many instances of persons who have tortured their friends, though I can not think of any at present. But that is no test by which to determine a devil, because devils are spoken of as believing in God, and trembling in view of God's power, therefore you ought to have asked whether we believe in God, and whether, also, we tremble in view of his power."

QUESTION.—"I think it will be admitted by all that the very last degree of depravity consists in a person's realizing delight, fun, or merriment in the torturing agonies of those who could but feel good will toward them. This is, and can only be, the test of a devilish nature. As to devils believing in God, I think they do so, because they as spirits are capable of comprehending better than men the immense power of God?"

Answer.—"As to the *last* test of a devil you have instanced, it only proves the race more corrupt than it appears to be. It can be demonstrated that all who leave the earth life in a wicked condition must come up into the atmosphere which surrounds the earth. Therefore to prove that we are just such characters as you have charged us with being, is only to prove the earth life more wicked than it is."

"It does not prove that the earth life is more wicked than it really is, but it proves that disembodied spirits may grow more and more depraved after leaving the earth life, or it may prove the existence of the original fallen angels."

"Yes, you are right, it does prove that disembodied spirits actually grow morally worse and worse, and it is because they have no good surrounding them. Evil, and only evil is the component element of this world. Wicked men in the earth life may seem comparatively good, but it is only because they are under restraint. There is not only no restraint here, but there is every thing to excite and provoke depravity, therefore they must necessarily grow worse and worse, because all the latent evil is excited and brought into activity."

QUESTION.—"But have not spirits the privilege of entering the church and hearing sermons, and have they not the privilege of remaining in the families of the pious and godly of earth?"

ANSWER.—"Very true, they have, and do so, and often criticise the sermon as you have heard us do, and we often come into pious families, and have the benefit of the heavenly atmosphere which surrounds them. But what does that avail when there is no protection? There may at the same time be hundreds of disembodied spirits, chattering and mocking you in the most hideous manner. The spirits in the air would no more suffer a fellow spirit to be pious and good than a herd of lions would allow a lamb in their midst. Besides they have no disposition to be good, have no affinity for good. This atmosphere of yours is the abode of all the accumulated wicked since the world began. Therefore, if you must know it, this place is as bad as wickedness can make it. There are no good people in this place. They immediately ascend, we think, to a higher region."

QUESTION.—"But why do you not go up there also?"

ANSWER.—"For this reason, we could not go up so far into the thin ether if we would, and we would not if we could, because of our moral antipathy to every thing that is good. You can but little realize how bitter is our antipathy to every thing good. You may say, then, why do we come around *you* when we admit you to be as good as any one can be in the earth life? Simply because you afford us some delight—you mitigate the tedious monotony of our life by your questions and arguments. We don't love you because of your goodness, but because you afford us delight. Your magnetism is softer than ours, and we delight to remain in it just as you like to remain in the sun in a cold day."

QUESTION.—"But why not go up, and remain in the soft and pleasant magnetism of the good above?"

ANSWER.—"Because, as I said, we can not if we would, for there is a geat gulf fixed between this place and heaven and no one can pass over it no more than you can pass to us in the body as you are?

QUESTION.—"Will not age, in time, make you invisible to those wicked spirits who have just emerged from the body?

ANSWER.—"I think not, because here are persons said to be a thousand years old, and they look just the same, only more intellectual, than those just coming from the earth life, just as a man looks more intellectual at forty than at twenty?

QUESTION.—"Have you any definite idea as to what will finally be the end of the wicked?

ANSWER.—"We have none. We only know what the scripture says about us, that we shall be burned up, that is destroyed. We believe its meaning to be having no conscious existance. Why we were made or what purpose we subserve is more than we can tell. There are a thousand theories among us which have come from earth life. But the jargon of ideas is worse than confusion confounded, as you very well know from experience. My own opinion is that they will all be destroyed at the end of the world. But when that will take place we know not.

QUESTION.—"But are not the doctrines taught generally by spiritualists denomiated in the scripture the doctrines of devils or demons?

ANSWER.—" Yes, they are, in very deed, the doctrine of devils or demons, because they generally reject the teaching of Jesus Christ and his apostles and followers. A. J. Davis was inspired to my certain knowledge by the prince of demons, or in other words the most intellectual demon belonging to the powers of the air. His Harmonial Phylosophy was all written under inspiration of demoniac influence. There is no Jesus Christ nor any doctrines taught by Jesus in his works—they are Christless or Anti-christ. The great central doctrine of the gospel, as you well know and as I learned in the earth life, is vicarious suffering or the suffering of the innocent for the guilty. It is seen in every plane and phase of earth life; the innocent bearing burdens for the guilty; the mother toils and wears herself out for her ungrateful children, who perhaps return only cursing for blessing and unrequited toil. Thus Jesus Christ has toiled, suffered, and given himself for the race, and for A. J. Davis, himself an ungrateful child, who turns and rends him by classifying him in his congress of spirits with some of the violent of the earth because the spirits giving it had so dictated it to him. There is not the smallest particle of the pure, self-sacrificing principle, called sacrificing self for the good of others, or in other words, the vicarious suffering principle, in all his works. They are purely the doctrines of devils or demons, and came from the pit of ignorance and villany. Spiritualism was conceived in sin and brought forth in iniquity. You may think strange in my saying so, but when I see the deceit, trickery and imposition palmed off upon the world as pure gospel truth it is enough to make all hell tremble. Some of us, if we have missed heaven, have not lost all interest in the success of the gospel, and for the following reasons: First, because every one saved and taken to heaven makes the number in hell less, and the fewer we have in hell the less of evil we have to bear with. Second, we have friends and relatives in the earth life and we feel interested to have them saved, even if we are not saved ourselves. Third, if we are wicked we should not be so wicked as to desire all others to be wicked.

"I was perfectly satisfied in witnessing your trials that there is a higher life in you, called in scripture or by the Savior, Jesus, the divine life or the eternal life, begotten by God himself in the soul. Therefore I can perceive that you have higher motives than spiritualists generally have. The inquiry by you was continually made to this effect: How is spiritualism to be used to promote the gospel or to develope the higher life, or the God principle in the soul? You were all the time looking for an advance of moral power in the gospel itself. The gospel had a certain degree of moral power when it made its commencement in the world. So you were expecting to find in the second gospel (and very properly too) a still higher degree of moral power. If spiritualism is to manifest that power in a higher degree, we know as yet that it has not manifested it at all. It is a dead carcass—a carcass that will be a stench to the good of the whole earth. Yes, bad as I am, I can see that its influence is evil, only evil, and that continually."

Thus, dear reader, ends a communication in some respects more peculiar than any which has preceeded it. It seems to be characterized by a degree of candor very remarkable for a soul that had missed heaven. Indeed, the author seems to posses a surprising knowledge of the pure spirituality of the gospel. Whether the one who dictated it to me obtained this spiritual insight into the gospel from being connected with those who had caused *me* so much suffering and deep soul experience of evil, I can not determine. All that I can know respecting the matter is the author's confession in these words: "I was perfectly satisfied by witnessing your trials that there was a higher life in you than in some others, called in the sacred scriptures the *divine life,* or eternal life begotten by God himself in the soul."

CHAPTER XVII.

The reader may be interested to hear my views respecting the divine or heavenly life in the soul. The views I shall advance will be characterized probably by being obtained more from my own deep soul-experience of the *higher life,* and the laws regulating it, than from any modern theology. Ever from my youth I was deeply penetrated with the fact that the gospel is the only hope of the world; that all the real, positive, good in the world is from the gospel, and that to crush it would be to destroy the only life-boat which has reached us from heavens bright shore. It must be evident to every rational mind that no man can or ought to believe in Jesus Christ, or in fact in *any thing*, without evidence.

The evidences which the gospel brings to us of its divine origin, and to compel belief in its system of truths, are the same kind as induces belief in any scientific fact, only the experiences are not derived from the natural senses but from the soul's consciousness. Man has moral sensibilities as well as those which are intellectual and physical; hence, soul experiences are as reliable as those experiences derived from the natural senses, and I think more so from the fact that the evidences of things derived through the five natural senses are liable to fail us, but soul experiences, derived through the agency of the moral sense or the moral sensibilities, can not fail us, for they are derived from God himself, who can not deceive. When it is declared that no man can call Jesus Christ Lord, or his Lord, but by the Holy Ghost, it is meant that the soul of man can not perceive the moral beauty, the exceeding loveliness of his character, but by the Holy Ghost. He alone can reveal him; that is, the moral excellency of Jesus in the soul which constitutes the Christian's hope of glory. Or when it is said that he that believeth on the son, hath ever lasting life, it undoubtedly conveys this idea: Jesus Christ is the brightness of the Father's glory—the express image of his person. Then to believe in Jesus Christ is to perceive or realize the fact of his divine excellence. This can only be effected by realizing the power of God, active in the soul, awakening it to consciousness, awakening its moral sensibilities and perceptions. So that before this revelation of God in the soul it may justly be said in the language of the prophet that Christ is to that soul as a root out of dry ground; but after he is quickened by the realization of God in the soul, Jesus is to him the one altogether lovely, the chief among ten thousand, or, in other words, he believes in Jesus, and this belief results from the evidences furnished by his moral consciousness, not from physical evidences derived through the senses but from soul experiences. To make this still plainer, it is recorded that Jesus upon a time, when addressing Peter, made this remark: "Peter, who do men say that I, the Son of Man, am?" Peter replied to his master in this language. "Some say that thou art Elias, others say that thou art Jeremiah or one of the prophets." Jesus said unto him, "whom say ye that I am?" Peter answered and said "thou art the Christ, the son of the living God." Jesus answered "blessed art thou, Peter, for flesh and blood hath not revealed this unto thee, but my Father who is in heaven." In other words, Peter had not obtained evidence of this fact through his physical senses but through his soul's consciousness, which was of God. Hence he could declare that Jesus was the Christ by evidences of a higher grade than could possibly be derived through the physical senses.

Man can not believe himself a sinner unless he has the evidences of it (by sinner I mean one who is conscious of sin and accountability to God.) The spirit of God alone can so enlighten and quicken the understanding, that the sinner perceives the spirituality of the divine law, perceives that the moral quality of every action is in the motive; hence the motive of the act controls the moral quality of the action. If a sinner is brought to a just appreciation of the fact that he is a sinner he must have the evidence of it by the power of that Holy Spirit, which is of God. Believing that he is a sinner implies a belief also in God as a moral governor. No man can feel a sense of guilt without he has evidence of it; his consciousness gives him the evidence that God exists and holds him accountable, and as the Holy Spirit can alone take off the things of God and show them unto him, therefore no man can have a just appreciation of God as a moral governor, and of himself as a sinner, without the evidence of it. And no power in heaven or earth can furnish the evidence to the soul of the sinner but the Holy Spirit. As no man can repent, truly repent, without this consciousness, therefore repentance is of God, through and by the agency agony of the Holy Spirit.

As Jesus says, when he, the spirit of truth is come, he will convince the world of sin, of righteousness, or what is right, and of judgment, that is of accountability to God as a moral governor. So then repentance, or the cleansing of the sanctuary of the soul, is the first and necessary step to the introduction of Jesus Christ into the soul as the hope of glory. Jesus Christ possesses in himself every divine perfection, being the brightness of the Father's glory and the embodiment of the *spirit of truth*. But the penitent sinner can not believe this fact until he has the evidence of it. This evidence he enjoys when the Holy Spirit opens his spiritual eyes, and reveals Jesus Christ to the soul, the one altogether lovely. Then he can say that Jesus is to him the *all* in *all;* and to be swallowed up in Jesus is to possess in himself every divine perfection. Thus it is verified that no man can call Jesus Christ, Lord, but by the power of the Holy Spirit. Therefore that no man can believe in Jesus Christ as his personal Savior without the evidence of it is clear, and as the Holy Spirit can alone furnish the evidence that enables him to believe, it must be perceptible to *all* that man's salvation depends entirely upon the proffered aid of the Holy Spirit.

This being the case, it must be a matter of the very highest moment to know the conditions upon which this proffered aid can be obtained. Jesus Christ, the Savior, has settled the matter in these words, "God, my Father, will give the Holy Spirit to him who asks for it." It is not given as the Greeks and many other nations believed as a kind of divine eflatus, which could be realized only by exalted minds—by poets and men of wisdom and eloquence, for Jesus declares that the Holy Spirit is given to *every one* that asketh. The conditions of his reception, therefore, are the inward desire for his presence and the act of asking. Then the fact of a higher, holier, even an endless life, depends upon complying with the conditions of that life, which are, first, the reception of the Holy Spirit; Second, a discovery of God as a moral governor; Third, the consciousness of moral impurity, arising from the fact that the motives and actions of the sinner have been wrong, neither the one nor the other having any reference to pleasing God as his moral governor; Fourth, the apprehension of Jesus Christ, as the embodiment and exempler of divine truth. Thus the sinner realizes in himself a lively faith in Jesus, which is the germ and motive power of an endless life.

Some have not apprehended the justice of God in his plan of salvation through Jesus Christ. But when we consider that Adam and Eve, the first pair, coming perfect from the hand of Omnipotence, possessed a perfect moral and intellectual constitution, and therefore were capable of rendering a perfect obedience under a perfect moral law; seeing then, that they had this perfect intellect, or perfect understanding to perceive their varied moral relations, and also a perfect conscience that enabled them to feel their moral obligations, therefore God could justly require of them a perfect obedience, and failing to render this, he would be justifiable in condemning them. The first parents, having violated the law, an impaired intellect followed, therefore their offspring also must have come into the world with an impaired or depraved intellectual and moral constitution. And as an imperfect being could not render perfect obedience to a perfect moral law, therefore the divine mind conceived and introduced the plan of making man's salvation not contingent upon his obedience, but rather upon using the *means* of *recovery* from the effects of the fall, or from the effects of Adam's transgressions, as it is said, "we are not under law but under grace."

This makes salvation, or recovery from the effects of the fall, to depend upon two considerations: First, a knowledge of the means to be used. Second, the ability to comply with the requirements of the conditions.

Man's nature being impaired or depraved, which depravity unfits him from clearly apprehending his moral relations, and from feeling as he should his moral obligations, therefore justice requires that man in his depraved condition should be aided by just so much of divine power as equals his depravity. This is supplied in the gospel plan of salvation through Jesus Christ. That plan contemplates the restoration of the race by enabling each one of them to render a perfect obedience to the law under which the race was originally created. And this is to be effected by bringing each individual member of the race under the power of the gospel.

This power, as we have already said, consists in the giving of so much of the Holy Spirit to each subject of grace as will enable him when it is received, first, to overcome the *repugnance* of his nature to humbling himself before God; Second, to ask God for the Holy Spirit; Third, the discovery to him, that the moral quality of every act lies not in the outward letter, but in the motive, and as a consequent to the discovery of God as the moral governor and of all sin as committed against him. Fourth, to the apprehension of Jesus

Christ as the one altogether lovely and morally excellent; as the perfect exampler of divine truth, and as the meritorious cause of all the blessings received under the gospel.

This being the divine plan of salvation by grace it must follow that the sinner's condemnation consists in *letting himself remain* in that condition of death or separation from spiritual life in which he is found by nature, and from which he actually refuse to be recovered, by willfully resisting the proffered aid of the Holy Spirit. This subjects him to an aggravated condemnation in the sight of God.

As Jesus said, there is a sin unto death, a sin which hath never forgiveness, neither in this world nor that which is to come. Peter addresses men in this language: "Ye stiffed necked and uncircumsized in heart and ears, ye do resist the Holy Ghost, as your fathers did so do ye." Paul expresses the same condemnation in this language: "If a man sin willfully (that is, if a man willfully and knowingly resists the influence which would lead him to a realization of the divine life) there remaineth no more sacrifice for sin," or no other way of salvation. In other words, if a man, after a full knowledge of the only way of salvation, shall willfully and deliberately reject it there can be no other hope left for him.

It may be asked, how we may know that the Holy Spirit is given to us; or in other words what are the evidences of the presence of the Holy Spirit in the soul? First, it *quickens the moral perceptions, giving a clear understanding of the soul's relation to the divine law*, and as a consequence, a consciousness of moral impurity. Second, it gives tenderness and susceptibility to the conscience and the moral feelings, so that the man is warned of the least approach of moral evil, and filled with dread of its consequences. Third, when not resisted in its operations it goes on to elevate the affections above earthly things, by causing the heavenly things to appear real and positive, thus giving to the soul an utter loathing of all moral impurities. Hence, the man who enjoys the presence of the Holy Spirit realizes that the kingdom of heaven "is not meat and drink," but righteousness and peace and joy in the Holy Ghost; or, in other words, the soul realizes that right doing invariably gives peace of mind, and as it is made conscious of rising to higher degrees of moral purity, putting on a higher and nobler style of manhood, the peace of mind which follows this conscious realization, hightens into joy unspeakable and full of glory. In the light of what has been said, it may be seen that the leading, central doctrine advanced by spiritualists, to-wit: That of the endless progression of all men to a state of holiness and moral purity, either in this world or the world to come, is false and totally at variance with reason and revelation.

First. Because we find, from the confessions and the grossly wicked conduct of these departed spirits, that death produces no change in the moral character of the man, but on the contrary whatever may have been his immoral habits, previous to physical death, they are essentially the same after that event takes place. So that the scripture is virtually fulfilled which says, "He that is filthy, let him be filthy still; and he that is unrighteous let be unrighteous still."

Second. There can be no Godly sorrow for sin where there is no gospel; no holy spirit, where every member of the society has his heart fully set in him to do evil.

Third. Because these wicked beings are at the present time in the air, or the earth's atmosphere, and are not in a hell of fire. Still they know not but that after the judgment it shall be said to them, "Depart, ye cursed, into everlasting fire, prepared for the devil and his angels." By their own declarations they have no more knowledge of the future than wicked men in the form, and the knowledge which they possess amounts to nothing, absolutely nothing, unless it be derived from the Bible, and *that they profess* to disbelieve. Hence the future to wicked men, whether in the body or out of it, is a positive blank.

The most probable conclusion that we can arrive at respecting the present state or place of the wicked dead, is this : It is the place of spiritual darkness spoken of by Peter and Jude, under which the wicked are reserved unto the judgment of the great day; and the place to which Judas, the betrayer of Christ, was consigned after his untimely departure from the world.

Fourth. From more than four years' converse with these wicked beings, and from hearing them talk to one another upon every subject that pertains to their present abode and social condition, I have, with much deliberation, arrived at these conclusions, viz: In the world of departed, wicked spirits (alluded to above), there is to be found essentially every variety of character to be found among wicked men in the earthly life. They virtually possess every element of character, except holiness, belonging to a human being. They are accustomed to the use of slang and cant phrazes of every grade and degree, from the obscene, debauch, and pest-house vulgarities, up to those eminently refined in phrazeology. But every expression emanating from their minds, and every imagination of their hearts is evil, only evil, and that continually. Many came to me professing great love to God, and very sanctimonious in their mode of expression, with words smoother than oil, yet a few pertinent questions enabled me to unmask them—revealing their gross hypocrisy and deceit, after which they would pass away with the remark, "You are getting too *sharp*." Some came professing to belong to the third or fourth heaven, and as a matter of course, pretended to be very pure and very happy, but these also were discovered afterwards to be gross deceivers, by acknowledging that they were engaged in torturing and deceiving me. They could not have belonged even to the highest grade of mind in hell.

Heaven is freedom from sin, and freedom from sin implies perfect obedience. Perfect obedience implies divine order and harmony, and only where divine order and harmony reigns can there be said to be heaven. There can not be degrees of heaven, any more then there can be degrees of justice, or degrees of truth. Justice is justice, and truth is truth, and there can not be degrees of justice nor degrees of truth. So there can not be degrees of heaven. The idea of heaven precludes the idea of degrees. All sin is a violation of law, and where law is violated there can be no heaven. If one sin were tolerated in heaven, then heaven would cease to be heaven. *Perfect bliss consists in the exercise of all the faculties of the soul upon their proper objects in a proper degree.* Therefore, if the faculties of the soul had insufficient exercise, or if they had not proper objects upon which the faculties should be exercised, or if they were exercised in an improper degree, then it must be evident that heavenly life would be but a transcript of the earthly life.

Here on earth is confusion, but in heaven perfect order reigns, and perfect order implies perfect purity. There may be different degrees of wisdom manifested in heavenly life, but there can not be degrees of moral purity. There are other considerations that might be brought forward to prove that the spiritualists' idea of degrees of heaven is absurd in the highest degree. It may be urged as an objection to the idea of heaven which we have advanced, that most Christians leave this world in an imperfect moral condition, and therefore they must enter heaven in the same moral condition in which they left earth. But this can not be, and for the following reasons: The Christian has the desire for moral purity, and earnestly strives after it, but in the earthly life he can not attain it, because his surroundings are so impure. The fountain of water issuing from the crystalized rock is pure at its source, and becomes contaminated only from flowing over impurities in the channel. So the Christian, having an innate desire for purity, but being prevented from its attainment by worldly surroundings, must, necessarily, when freed from his impure surroundings, be absolutely pure; and if morally pure, then a fit subject of heaven.

Many worldly persons desire to be happy, but that is a matter very different from a desire to be morally pure. The true Christian has the leaven of righteousness within him; he hungers and thirsts after it, and this element of his renewed nature constitutes the essential difference between the righteous and the wicked. All Christians have not this desire for moral purity in the same degree; but all have it to some extent, or they would not be the chosen of God. He that hath this hope within him, purifieth himself, even as God is pure. Then the doctrines of spiritualists at the present day are false, and without foundation in reason, or the revelation of Jesus Christ, and should not for a moment be entertained by Christians.

Besides the doctrine of the endless progression of all men, to a state of holiness and moral purity, there is another doctrine equally pernicious to the welfare of the world. It is asserted that sin is but the soul's experience of that which is unfit and improper. This is, to all intents and purposes, false. It is false because not in accordance with the facts of nature, and in the record made by Jesus Christ. Jesus said sin is the violation of law. Law is the divine order or arrangement in nature, both in respect to the physical as well as the moral world; and all violation of law results in injury to the person so violating, whether of a physical or a moral nature, as the case may be. So the idea of indulging in sin to experience its unfitness in our bodies or minds, and by so doing learn to avoid it, is as absurd as the idea of taking a little deadly poison to ascertain its effect upon our physical system.

CHAPTER XVIII.

All must have observed the power of habit in controlling the activities of the mind and body. When I speak of habit, I mean that tendency of our nature by which any given act, physical, mental, or moral, often repeated, is rendered easier to be performed. Then the repetition of a sinful *act* enables the performer to do it easier and with greater pleasure to himself, therefore the tendency of sin or violating of law is to bind its subjects with fetters of brass. Paul says, "His servants ye are, whom ye obey, whether of sin unto death; or righteousness unto life." The evident meaning of which is, that sin often repeated makes the man who performs the act a servant, or slave of sin; while the man who does righteousness becomes by this same law, or tendency of our nature, confirmed in rightdoing. So it must be evident, both from nature, and from revelation, that indulgence in sin can never reform the man who practices it.

We have all inherited depraved natures—impure, unholy, and opposite to that which is good. This is not always manifested at once, but lies latent in man's nature, and as occasion may come to arouse it, stir it up and draw it forth, it appears in all its power. Thus it was with Hazial the Assyrian, who said to Elisha the prophet, "*What*, is thy servant *a dog*, that he should do this *thing?*" when it afterwards appeared that he was guilty of doing the very deeds that his nature then revolted at. So then all men are naturally averse to holiness, from a tendency inherited in their very being, and this can be overcome only by the power of God. This natural repugnance is not the result of education, for it so happens that minister's sons and daughters are overcome by it, and run into sin as naturally as others who may have been raised under circumstances the very opposite. Then it must be evident that the whole doctrine of indulging in sin, to give the soul an experience of its effects that it may gain a repugnance to it, is false and untrue to nature. This is so common and evident that it needs no further proof to make it manifest. It is the order of nature for men to follow that which they love, and if restrained for a time they will return to it, as the dog to his vomit, and the swine to his wallowing in the mire. The definition of a saint is one who would do no evil even if left unrestrained. So the the truest definition of a sinner is one who will do wrong whenever not restrained by the laws and usages of society.

An allusion was made, in the preceding communication, to the effect that these beings above and around me were acquainted with my thoughts. Reader, this may seem strange to you, but not at all strange to me. I have noticed, through several years of careful watching, that those invisible intelligences, which the Bible calls demons or devils, are not only acquainted with my thoughts, but with my every feeling and emotion. When in rapport with me, as they call it, they actually seem to be a part of my being. Their thoughts appear to flow into my mind, many times easier than can the thoughts of men in the form reach my mind through the medium of language. My every thought is as transparent to them as to myself; and this very fact is one source of very great annoyance to me. Most persons, I think, let their thoughts run pretty much as they please. But since I have been brought into this strangly impressible condition, I am obliged to be as careful of my thoughts as ordinary persons would be of their words while in the company of others. No person can have any idea of what the *luxury* of being alone is until they are deprived of it. I am now in no sense ever alone. My thoughts to them are perfectly evident, and lead them to indulge in the sharpest critcisms. Whether they were always so, I can not tell; or whether the thoughts of other people are known to these beings of the air is yet, to me, unknown. But the fact that they know my every thought, I know to be true. This, to me, has been a fiery ordeal—an ordeal that has tried to the utmost my very sensitive nature. For example, if I should choose to indulge in *thoughts of censure* or repoof towards any person, some one of the ever present intelligences would say; " Be careful—remember your own frailties. " And another then would remark: " No, talk to him sharply; he deserves a severe rebuke. " My mind would then perhaps pass on to some other thought, leaving them to finish the discussion. Frequently as the tenor of my thoughts did not appear to please them, I would remark by saying, " well, if my thoughts do not please you, perhaps you can suggest some better topics of thought. " To this they replied, " Just now we are not getting up topics of thought. You will excuse us if we follow your train of thought, only allow us to criticise and make such remarks as we please." Then they would perhaps continue by saying, "You are a Christian, and of course your thoughts are, or ought to be good." To this I would reply, "Yes, they ought to be good, and much better than they are, but I most earnestly desire that you would be silent witnesses of my thoughts, rather than unmerciful critics." To this they replied again, "We are great critics, for the reason that we are always at it." So in this, and many other ways, I have tested the fact that these invisible critics are perfectly acquainted with my every thought. Nothing can be hid, not even the slightest shade of a thought. They seem to be always present, always gazing right down into my inmost soul. Their continued presence would at times become an intolerable burden. Often I would request them to leave me, if it were only for a few moments; but alas, even that was denied to me. As a cat watches her prey with a most vigilant eye, so these strange beings would apparently watch my every motion and thought. But as I have said I became convinced in many different ways, that they were perfectly acquainted with my thoughts, so if acquainted with *my thoughts,* why are they not acquainted with the thoughts of others, and if that be a fact that they are as perfectly acquainted with the thoughts of others as my own, then what an immense power they are capable of exerting over the world of mind! This gives additional proof that they are the legions of the prince of the power of the air working in the hearts of the children of disobedience. Many incidents of my early childhood have they repeatedly taken from my mind, and then have made the attempt to bring them in as proof that they were my old acquaintances or relations. But a few questions or a few inquiries soon satisfied me that they were not the persons they represented themselves to be.

It must appear evident to all that with such abilities they are capable of practicing almost any amount of deception, especially upon *mediums,* and upon those who inquire of them with reference to the condition of their friends or relations in the spirit world. For instance, a person calls upon a writing or impressible medium, to ascertain with regard to a deceased friend or relative, or with refference to the same person being yet in the form, but absent. Knowing, as they do, the thoughts and conjectures of the person calling, they are capable of making up from the thoughts of their own minds a very plausible story. Besides, I think, they can pass with great rapidity from place to place, and must necessarily become as well acquainted with the affairs of every individual of a nation, as any ordinary person can become acquainted with the affairs of a single neighborhood, and perhaps more so.

I mention this that persons may understand and know to what extent these beings can impose upon those in the form, if so disposed. I will call the reader's attention to the reply to my question in reference to the teachings of spiritualists, as being in fact what are called in the Scripture the doctrines of devils. The first dash in the reply is in these words:

"They are in very deed the doctrines of devils, or demons, because they generally reject the teachings of Jesus Christ and his apostles, and follow the teachings of A. J. Davis, who was inspired, to my certain knowledge, by the prince of the *demons;* or in other words, the most intellectual demon belonging to the powers of the air." The spirit goes on to state, that the whole of Davis' Harmonial Philosophy was written under demoniac influence, and says in these very significant words, "There is no Jusus Christ, nor any doctrines taught by Jesus Christ in his works; they are *Christless* or antichrist."

Of A. J. Davis and his works I know but ver little, especially his Harmonial Philosophy. But what I know of his teachings, I have heard from Trance Lectures, and the followers of his system, whom I have met in the daily walks of life. I would say in reference to them, that they invariably reject the Bible as the word of God, and Jesus Christ and his whole system of gospel religion. They will hardly admit that Jesus Christ was even a good man, because they say that he must have been conversant with spirits of considerable wisdom, and if so he must have deceived the people; intentionally deceived them, especially with reference to the progression of all men to a state of final holiness, either in this world or that to come, for he positively taught the endless perdition of the wicked in the world to come. His deceptions are the more evident, they say, because after his death he had opportunity to prove, by actual observation in the spirit world, the positive falsehood he had uttered or been impressed to utter by wicked spirits in his earth life. But instead of recanting what he had said, he actually inspired Paul, John, and others to utter the very same doctrines in their writings.

I once heard a trance speaker, a lady, utter these words, "Jesus Christ taught many truths, but his system of religion called Christianity will be overthrown, utterly overthrown, especially the doctrine taught by him of the endless punishment of the wicked and the doctrine of vicarious suffering; the doctrine of the Holy Spirit's power and influence; the doctrine of the fall and depravity of all men, and the doctrine of satanic influence."

These doctrines, and some others, I have found particularly repugnant to spiritualists. Said I to a leading spiritualist:

"If A. Jackson Davis asserts one doctrine as positively true, and Jesus Christ asserts as positively to the contrary, how shall we decide who is right or who is wrong?" Said he,

"Believe that which is the most reasonable."

"But," said I, "reason has nothing to do with it. Reason is confined altogether to the domain of facts. It arrives at just conclusions, by collecting all the facts bearing on the case, and in no other way. Now suppose one witness to assert that the wicked are punished by being banished from the presence of God, and the glory of His power forever; and the other witness, equally as creditable, asserts that they are not punished at all, but immediately go to a place where they enjoy as much of God as their natures are susceptible of enjoying, then how can *reason* decide, the testimony being equally balanced?"

"Well," said he, "it would be hard to decide if they were both equally creditable witnesses."

"But," said I, "if one witness should positively assert that he was taught of God, who could not err, and as a proof of the fact, should calm instantly a violent tempest; raise the dead; restore perfectly a withered hand, and do many other wonderful works; and the other witness should assert that he was taught by the spirits who came from the earth life, and should make no pretensions to working miracles as proof of his divine mission, how then would the case stand?"

But he was silent.

Another assertion in the communication struck me very forcibly. It was in these words:

"We feel interested, in more ways than one, in the success of the gospel. First, because that every one saved and taken to Heaven makes the number in this dread abode less. Second, we have friends and relatives in the earth life, and feel interested to have them saved, even if we *ourselves are lost.* Third, we are not so wicked as to desire all others to be wicked."

This language is in spirit precisely the language of the *rich man* in hell, of whom the Savior speaks. I can not but think the writer to be sincere, that he speaks the convictions of a mind fully convinced in reference to hell. I have often heard them say, (after torturing me for several hours by *epithets, insinuations,* sarcasms, and the foulest and most obscene expressions), that they (that is themselves) were too mean to live, that they were not *fit* to live, that they ought to be blotted out. I think, as far as I could judge from the conversation of those around me and their conduct towards me, that a part of the sufferings of hell consists in an ever restless *desire* to do something, and that something that they want to do is that which will sink them lower and lower in their own estimation, and lower and lower in the scale of moral being.

Another idea brought to view in this communication is in these words: "I was perfectly satisfied in witnessing your trials that there was a higher life in you, called by Jesus Christ, the divine life or the eternal life. And as a consequence of this I perceive that you are actuated by higher motives than are many of the so-called spiritualists. Your inquiry," says he, "was continually to this effect: How can spiritualism be used to promote the gospel or to develop the higher life, or the God principle in the soul? You were looking for an advance of moral power in the gospel itself. You say that the gospel of Christ had a certain amount of moral power at its commencement. So spiritualism, if it is to be the second gospel, must have a still greater moral power." It is then asserted in these words, "We know that spiritualism has not as yet manifested any *moral power* at all. It is a dead carcass, a carcass that will be a stench in the nostrils of all the good of the earth. Yes, bad as I am, I can see that its influence is evil, only evil, and that continually." Now this expresses in full the real burden of my soul. I was sure that I realized higher life in God than any spiritualist with whom I had been conversant. And I think I say the truth, when I say that no spiritualist, as such, leads a life of earnest prayer before God. Some of them, perhaps, were men of secret prayer previous to their becoming spiritualists and have continued the practice. But the thought I design to impress is that spiritualism of itself never led a man to realize the higher or eternal life in God, and as a natural consequence of that life be led to *call upon* God in secret prayer. On the contrary, it leads men away from the higher life, by destroying the idea of sin as being committed against God as a righteous moral governor; by asserting that God is a principle, an all prevading principle, but not a personal, intelligent being; by asserting that sin is the soul's experience of that which is unfit and improper, and as soon as the unfitness is asertained the person so experiencing will forsake it.

It must be evident that this assertion is not true. Every man in his natural condition is antagonistic to that which is holy. He avoids recognizing God as having any authority over him, much less that sin is committed against him. Such refuse to entertain God in their thoughts—that is, a God to whom they must render a strict account. Also the natural course and tendency of sin, as exhibited in the life and in the daily act, is to bind its victim more firmly as each sinful act is repeated. Also upon the recurrence of each sinful act it becomes still easier to be performed. For instance, who does not know that the indulgence of anger increases its power over the one who yields to it, or the indulgence of profanity increases the power of that worst of all habits, over its victim. Take the case of lying. Every child knows that telling untruths gains power by repetition until the habit is confirmed in its fullest force; the man, (unless renewed by divine grace) goes down to his grave a confirmed liar, and men say of him, the "ruling passion is strong in death," or "that which is bred in the bone stays long in the flesh."

It is useless for me to attempt to prove further that spiritualism has never manifested moral power to lead any man to realize the eternal life of God in the soul. On the contrary, the man indulging in sinful habits finds that the indulgence gives him less and less power, and less inclination to reform. The cause of this is depravity; physical, intellectual and moral depravity. The man's whole nature is impaired, and this he inherits. The lust that burned in the nature of the father is felt in the fullest force in the son. The ungovernable temper that is seen and felt in the parent is shown in the child, in the youth, and in mature life. The propensity of lying in the parent is witnessed in the child, and often provokes the familiar saying: "A chip out of the same tree." Moral infirmities are transmitted from parent to child, as well as those which are physical. As a dead carcass has no power to *arrest its decomposition*, so man has no power to arrest the ruling force of sin after having indulged in it, or after having become the servant of sin. For this reason there is an actual necessity for grace, for the application of a moral power above and beyond him, and that power is known as the Holy Spirit power. Spiritualism recognizes no such divine power; it is indeed a dead carcass. It finds man a slave to sinful habits, and leaves him as powerless as it found him.

I will notice another point which I consider germinal of the first importance, as it helps to more fully understand the way of eternal or higher life, which is of God. The point I wish to notice is expressed in these words: "No man can feel a sense of *guilt* without he feels or realizes that God is the *great moral governor*, who holds him accountable for every thought, word and deed." There are but three sources of suffering in the world, viz: Physical suffering, which is occasioned by the violation of physical laws; mental suffering, which is occasioned by losses and disappointments of every kind; and moral suffering, which results from the lashings of a guilty conscience. The first kind of suffering can be mitigated by obeying, as far as possible, the laws of physical health. The second can be alleviated only by enacting such laws and regulations in society as will secure as far as possible against losses; but moral suffering or guilt can be relieved only by complying with the will of the *moral governor*. That is, the offender must, to restore peace of mind, either bear the penalty of the law, or comply with the condition of pardon and forgiveness.

Men of all ages have resorted to three methods to relieve themselves from a sense of guilt, or remorse of conscience. The first class resort to the method of annihilating God; that is, they bring up in array before their minds all the facts that can be collected to prove the non-existence of the creator or moral governor, and shut out from their minds all other conclusions. Thus by a mighty struggle to stultify their reasoning faculties, and close their minds to the facts of nature, they perhaps obtain at times peace and quietness of mind or a release from the goadings of conscience. They are called Atheists.

The second class resort to the method of annihilating man. They collect facts and attempt to prove as *dieth* the *beast* so dieth man. Like the Saducees of old, they declare against the existence of either angel or spirit. By so doing they relieve themselves from all sense of accountability, and thereby obtain peace of mind and a quiet conscience.

The third class acknowledge the moral governor, and the accountability of man, and endeavor to ascertain the conditions of pardon or forgiveness, and by complying with these conditions obtain a lasting and permanent peace, permanent, because founded on the facts of nature. This class are called Christians. Not all Christians obtain a permanent peace, because they do not, like Paul, labor to have a conscience void of offense toward God and man. But those who do walk not after the flesh to fulfill the lusts of the flesh, but after the spirit, have not only peace but joy unspeakable, and full of glory. This results from the conscious love of God, which flows into the soul as the evidence that he approbates their conduct.

Spiritualists annihilate God by declaring that he is a principle, a law which pervades all nature, and like the attraction of cohesion or gravity, has no moral feelings. He neither approbates or disapprobates, in fact they reject altogether the idea of God as a personal being, an Infinite Father, filled with love and compassion. To realize God as an infinite, loving Father is *eternal life* itself. It fairly makes my soul thrill with joy unspeakable to realize it. O, the hight, the depth, and the breadth of the love of God, in Christ Jesus, our Lord.

One principal object that I have in view in writing out and giving my experience to the world, is this: Experience is knowledge, and knowledge, positive knowledge, is of priceless value. To give the reader all of the facts of my experience with the powers of the air would be literally impossible. To give all that they said in my hearing, and all that they did to my person would fill a very large volume, and present you such a picture of the moral condition of the wicked dead as would make all nature shudder at the thought. For as one of the inhabitants of that place said to me more than once: "There is a place where all the sewers of God's moral creation terminate. A place more awful than the mind of man has conceived of. It is not a place of perpetual fire, of lurid burning, for that has no tortures to spirits—they can exist in the fire as in summer day—but it is a place from which all that is pure and good is banished. Your world has a commingling of good and bad; but ours has no good at all."

CHAPTER XVIII.

I will now present the reader with another communication in answer to the following question:

"Why was the Holy Spirit manifested in greater power under the Mosaic dispensation than under that of the Patriarchal, and why greater under the Christian than under that of Moses?"

"We answer, that the race of man is progressing not only to a higher degree of physical sensitiveness, but also to a higher degree of moral and intellectual sensitiveness. This being the case, it must be evident that the more sensitive the nature the more decided and positive will be the manifestation of the divine influences. That man is a progressive being, and that his whole nature is becoming more and more sensitive, can be made evident from the fact that all nature is putting on higher and nobler forms of beauty. Every kind of shrub or fruit tree, and all kinds of animal life put on a higher degree of beauty or a larger and nobler form as the soils of the earth become more refined, and the nourishments of animal and vegetable life become purer or are furnished in greater abundance.

"The history of man itself furnishes abundant evidence of the position which we wish to establish. Go back in the history of the world to the period when history first begins its record. Does it not speak of man as a gross physical being, wholly under the influence of his passions? Does it not speak of there being giants in those days, men of renown, mighty men—not intellectually mighty, but men of immense physical strength? Does it not declare that the earth was filled with violence? (Meaning undoubtedly that these men were wholly absorbed in the, to them, all-important work of destroying their fellow men because they had the physical power to do it.) Afterwards we read of men who were famous on the page of history for their intellectual powers, men who spent their whole time in either hearing or telling something new. These men became famous for their wisdom, eloquence, research, and erudition. They founded seats of learning and made them famous, the world over, for the ripeness and proficiency of the scholars which graduated from them. They laid the foundations for classic lore, and to such an extent did they carry their researches into every branch of human learning that their works have been the standards of classic elegance throughout all subsequent time. The intellectual giants of the world flourished from before Christ, five hundred years, to about the same time subsequent to his advent.

"About this time, Jesus and his apostles introduced the moral element, which had been seen only as a mere germinal principle during the Patriarchal and Mosaic ages, but under their fostering care became a mighty tree whose leaves were for the healing of all nations. The present moral aspects of the world are but the beginnings of that grand and glorious age which will prove to be the crowning glory of the world. Hence, the history of the race shows that man has progressed, first in the direction of the physical, then the intellectual, and finally the moral. All this has been brought about by the operations of nature, by refining the soils, by furnishing a purer and better aliment for the nourishment of both plants and animals.

"You will discover from the inferences which may be drawn from what has been shown to be the truth in reference to the progressive development of the race, the reasons why the Holy Spirit was manifested in greater power under the Christian dispensation, than under the Mosaic, and also, in greater power under the dispensation of Moses than under the dispensation of the patriarchs. You must discover that the culture of the physical, intellectual, and moral natures of man only renders him the more sensitive, and hence the more sensitive the nature the more easily influenced by the pure and holy beings who have, during all ages, helped to the extent of their ability, to educate and improve the race.

"Now it must be evident that the call of Abraham—the sojourn in Egypt and finally their call to take their place among the nations of the earth, is only a part of that grand plan which will eventuate in the entire redemption of the race. The Jews were hedged about or fenced in as a nation so as much as possible to keep them aloof and separate from all other peoples. Fasting or intervals of abstinence from all food was carefully enjoined. The kind and quality of food was carefully prescribed. The beasts were separated into clean and unclean, and particular directions given to use for food only those animals denominated clean. Polygamy was proscribed and the marriage relation carefully guarded. Many kinds of washings and ablutions were imposed, and the intervals of time during which they were to labor and rest carefully pointed out. All this showed that the physical development of that nation was as carefully attended to as was that of their moral or intellectual natures.

"Notice the results of this careful attention to the laws of their physical well being. Hardly a year had elapsed, from the time of their exit from Egypt, before we read that over seventy of the common people possessed the gift of prophesy. So elated was Moses at this manifestation of the divine favor, that he cried out in the exuberant extacy of his soul, Would to God that all the Lord's people were prophets, and that the Lord would put his spirit upon them., How much the abstinence from all animal food and the fact of their feeding on the doubly refined food called manna had to do with this prophesying, I leave you to judge. The truth is, it had much to do with it. If you knew how difficult it is for us, coarse and posative as we are, to so come in rapport with your physical nature that we can control, even your mind as at this moment, you could then in some measure comprehend how difficult it must have been for those refined and holy spirits from the Lord to have controlled the seventy, even under the most favorable circumstances. If *we are obliged* at times to double and thribble our power a hundred fold by combinations upon you, what a mighty host must have been employed to control those seventy! These expressions in your Bible, 'Come with the clouds of heaven,' 'Behold he cometh with clouds,' 'The Son of Man shall come with ALL *his holy* angels,' clearly indicate what immense numbers of the exceedingly refined, and pure beings of heaven it takes to produce conditions necessary to control even the best and holiest of men. The law in the case is simply this: The more refined in every way the man, the less spirit power it takes to control him; but the more coarse and positive the nature, the greater must be the amount of spirit power employed.

"You will discover the answer to your question, 'Why was the Holy Spirit power manifested in a greater degree in the day's of Christ, than in the day's of Moses, who lived nearly two thousand year's before that time?' to be this: The nation, as a nation, had, by the special physical, intellectual, and moral discipline through which for centuries it had passed been prepared for this greater display of the divine power, and because, Jesus, understood the laws of spirit control better than any and all others who had preceded him, he well knew that oneness of purpose on the part of those receiving the divine gift, and also, the oneness of their desires and their physical and moral purity, were all very essential conditions, in order to there being displayed any great amount of the divine power. Your revival preachers who have met with any very great success, have carefully enjoined the conditions which Jesus prescribed."

QUESTION.—"How did you come in possession of all this knowledge respecting the laws of divine manifestation, and how is it possible for you to understand how the divine spirit manifests himself, and still be in opposition to that spirit, and all else that is good?"

ANSWER.—"I am in possession of this

knowledge just as you come in possession of knowledge, by observation of the laws of nature; by converse with other minds, and from my knowledge of the nature, design of and intention of scripture rvelations. You may think, because our natures are not in sympathy with good, that therefore we are ignorant of the laws of divine good. But it is not so. The very fact that I have missed reaching heaven, has stimulated my own mind to look into the reasons of my great blunder."

QUESTION.—"But do you not feel very sad over the fact of your lost condition?"

ANSWER.—"I did at one time, but I am now perfectly reckless. I have no concern about the future."

QUESTION.—"Are you one of those who laid the plan of deceiving me, and have given to me so much pain?"

ANSWER.—"You forget that you are passing through a strange course of treatment; which may seem to you to be cruel, and it is so when you consider that it is for the amusement of wicked beings around you; but when you consider it in another light it is different. You should remember that God may have a purpose in permitting all this cruel treatment. He can make all things work for good."

Question.—"But God permits *very many abuses* in this *world*, which he expects the sufferers to find out the cause of and arrest?"

Answer.—"Very true, you have long been afflicted, and will be much longer; but as you have not found out the cause, or applied the remedy, therefore either you or God is to blame for it."

Question.—"This outrageous abuse can originate from no good source, God has ordained that each of his creatures shall be entitled *to his own person*, to his own distinct individuality, and the extreme cruelty that has been received can not be of God."

Answer.—"You say this 'outrageous abuse,' as if all did not know that abuse of any kind is 'outrageous.' God has ordained that all shall enjoy certain rights, but they are not always respected?"

Question.—"True, and that is the reason

that God has no special design in permitting this abuse, though he may overrule it, to accomplish some great good. You intimate that God has a purpose in subjecting one of his creatures to the unmitigated and confessed abuse of others?"

ANSWER.—"Very true, it is abuse of the very worst type, but God permits it, or it would not be so, but not that he is pleased with it. You must remember that many abuses are allowed but not sanctioned."

QUESTION.—"Then why do you continue the cruel treatment, if you think God does not sanction it?"

ANSWER—."Because we have nothing better to do. We delight in tormenting any one, especially those whom God loves as we think he does you."

QUESTION.—"Then you indulge in pure malice without provocation?"

ANSWER.—"Very true, we are pure malice itself."

QUESTION.—"But that is attributed only to the devil."

ANSWER.—"Very true it is attributed to

the devil, and all who are like him. We are like the devil of your scriptures, and are so because God made us so. We are for a purpose or God would not have brought us into being."

QUESTION.—"But you made yourselves what you are, did you not?"

ANSWER.—"No, we were born into the earth life, without our knowledge or consent; placed under laws which we but imperfectly understood,—which we could not keep, and are then condemned for their violation."

QUESTION.—"But did you not enjoy the means of obtaining eternal life through the gospel of Jesus Christ?"

ANSWER.—"Very true we did, but failed to avail ourselves of the means."

QUESTION.—"Then you remain vile because you failed to avail youself of the means of recovering from that vileness?"

ANSWER.—"Very true, it is by neglecting to act that we are in this condition, but were we not born with the inclination?"

"A man may be born with a physical or moral infirmity and may greatly aggravate that infirmity, or may use the means of recovering from it."

"Very true, we have greatly aggravated our moral infirmities, and of course God is not chargeable with that."

"While an infant you were comparatively a pure living force; but instead of improving your nature, imperfect though it may have been, you spent a life time in injuring it."

"Very true we did, but the conditions of society helped very much to make us bad."

"Not necessarily so."

"No, not necessarily, because man could, in a measure, banish himself from society."

"But do you never expect to be better?"

"Never. We are the debris of God's moral creation, cast off as far as we know only to be destroyed."

"Would you not, like all other beings, feel happier in doing justly, loving mercy, and walking humbly before God?"

"No; not at the expense of doing right. If we were to do right, we should be forced to leave you."

"But no creature can be as happy in doing wrong as in doing right, and certainly you desire to be happy?"

"Yes, I do; but I can be happy in doing wrong."

"Do you not believe there is a God?"

"Yes, we believe in God, and in Jesus Christ, just as we did in the earth life, but we can not love him as you do. We believe that he is willing we should do wrong or he would prevent us from doing it."

"But he has positively expressed his will, both in the book of nature and of grace, that his creatures of every grade and class should obey the laws under which they are created."

"Very true, but we are depraved, were born so, I think, and can no more keep the moral laws, as you call them, than can the hyena of the forest."

"Are you always actuated by feelings of anger, malice, revenge, hatred and other malevolent feelings?"

"By no means. We never show anger unless provoked, as you know by experience."

"Then if you can keep from being angry for a short time why not for a longer time?"

"Simply because we are not provoked. If we were never provoked by any one, I doubt whether we should ever be angry; and so if we had no opportunity of doing wrong, of injuring others, we should never be guilty of any wickedness at all."

"But you say that you believe in God. If so you can pray to him where you are for grace and strength to do his will, just as in the earth life."

"By no means. We are surrounded by unmixed evil, for us to pray here would be the same as for you to pray in the midst of ravenous beasts, or more properly in an uproarious, low-flung theater. Even if we had, and could feel the spirit of prayer, and should lift up a single petition to the throne of God's grace for help we should be instantly surrounded by a legion of chattering fiends who would torment the life out of us if it were possible."

"You say that you are surrounded by unmixed evil. As you inhabitants of the air surround us, are we not surrounded by unmixed evil?"

"Not at all; your natures are but in a degree sensitive to our influence. Besides you are told in your scriptures that you are kept by the mighty power of God. That to me indicates a special providence."

" But why are you not willing that *I* should pray, and why do you interrupt my prayers by strange psychological images?"

"That does not come from any act of my own. It is produced by others more wicked than myself, who have a perfect hatred of prayer. I have sometimes indulged a little that way, but not often. I generally love to hear you pray, not because I join in, or have even a desire to do so, but because you seem to realize so much of what I should think would constitute the bliss of heaven, viz: The love of God shed abroad in the heart."

" What do you suppose is the reason that I enjoy a higher degree of the divine presence now than formerly?"

"I think because your nature is vastly more sensitive, and that angels or something divine surrounds you, though we can see nothing."

"But do you not hate pious people, and is that not the reason why I have been so illy treated by you?"

"I do not hate good people, as such, but we all have a perfect hatred of *pious* people, because piety indicates to us that God has special favorites or pets, and as Joseph's brethren hated him because of his father's partiality, so we hate them on account, as we think, of God's partiality."

"If pious people have used the means of grace, and by so doing have been elevated to a high and holy place in the scale of being, ought they to be hated for it?"

"Not at all. We don't hate them for what they may have done, but we hate them because we believe God loves them, and has shown them special favors and denied them to us."

"But did he not make you the same offers of grace while in the earth life that he made to others?"

"Yes, the same. I well remember that I enjoyed many offers of salvation through Jesus Christ, but rejected them all; yes, deliberately rejected them. I remember times when I must have been favored with the striving of the spirit, but went to places of amusement and wore my impressions off, and as a consequence I am here in what the Bible calls chains of darkness."

"But have you no hope that God will yet have mercy upon you?"

"We have no assurance that would amount to a hope, for hope is made up of desire and expectation; we desire to possess a better nature, but have no expectation of obtaining it. We try to believe the theories so prevalent here that in the course of time we may have our natures changed, but they are so transparently false that I can not believe a word of them, for I very well know what the Bible says about the final end of the wicked."

"But could you not enjoy heaven with your present nature, if permitted to go there?"

"No; heaven to me, with the nature I now have, would be a double hell. I sometimes think it a mercy in God that he lets me remain where I am, but a mercy *I don't thank him for*. I hate God because he made me with a nature to hate that which is good, and pure, and cleave to that which is evil. I often wonder what I *was* made for."

"But you did not hate the pious and the godly in the earth life?"

"Yes, I had the same antipathy to them then that I have now, only then appeared the *germ*, but now appears the full grown stalk, ladened with the fully ripe seed of hatred."

"But do not the pious dead surround those who are still in the body as guardians from the influences of evil?"

"They are never seen by us if they do. We see nothing around the pious, any more than around the wicked. But we are often around them ourselves, infusing into their minds some infidel or atheistic thought to see how they will receive it. We take delight in disturbing and irritating them, just as we do you."

"But do you not often cause them to fall away, and relapse into wickedness, and by so doing increase the number of the lost?"

"We think not. The really pious seem to have an understanding that the impressions of evil are from the wicked one, and so instantly reject them. We have no desire to increase the number in the world of suffering and wretchedness."

"But do you not think the angels minister to the pious in the earth life?"

"Perhaps they do, but we can not see them any more than you can. They are as perfectly invisible to us as they are to you. That they might reveal themselves is possible, but we never saw one."

"How do the inhabitants of your world mostly spend their time?"

"We spend the time mostly since the discovery of the mediumistic communications in developing mediums; in making psychological experiments with them, and in communicating through them."

"Do you not think that good spirits develope mediums, and communicate through them as well as yourselves?"

"I think not, and for this reason: The mediums are all developed by the leading go-ahead spirits. Their magnetism being sharper, stronger and more positive, it is therefore accomplished by them in a shorter period of time. That they communicate through them, I think is not the case, and for these reasons: First, the saints, as we call the pious (but in derision), are too refined, we think, to make anything but a slight impression on even the most susceptible in the form. Second, even our power is so limited that there are but very few mediums who can be impressed singly and alone; it often requires scores and hundreds to create such conditions as will enable the operator to impress the medium. Third, there are but few pious mediums in the world that we known of, and it is not at all likely that holy beings, even by combination, could control them to write or speak. Fourth, there are no mediums to my knowledge who have any of the soft, agreeable magnetism about them which we think comes from the Holy Spirits but yourself, and you certainly know that no good spirits have ever controlled you, therefore we think we are warranted in the conclusion that no pious dead, nor the spirits of just men made perfect, nor angels have any thing to do with controlling mediums at the present day."

"But how is it, then, that many of them receive messages purporting to come from the apostles and other eminently good men, messages that seem to breathe the very temper and spirit of Christ himself?"

"Because, in this world, which, as I have said before, embraces the atmosphere of the earth, there are very many eminent ministers of the gospel, though *selfish, wicked men*, who in the earth life were noted for their piety and learning, and for their eloquence, too. Besides, there are men of the most exalted talents, intellectually, that the world ever saw, existing here in this place, and they often communicate through the various mediums, possessing all the talents they had in the earth life, not in the least impaired. They are capable of preaching as angels of light, and would deceive, if it were possible, the very elect. Hence, you received those first communications from persons of that class."

"How is it that these wicked spirits can depict, with such vivid, life-like representations the ineffable glories of the heavenly world?"

"How did they do it in the earth life? They evidently did it by the aid of the most wonderful of all faculties, the imagination. In the same way they build up theories of phylosophy. There has been, to my certain knowledge, as many as a score of different thoeries respecting the origin of the world, and the human race, communicated through mediums to the earth life."

"Are all nations of the earth alike impressible, and consequently alike exposed to the deadly and corrupting influences of spirits in the air?"

"No, the nations the most susceptible to our influence are the Anglo Saxon, the Celtic and the Goths and Visgoths of the North of Europe, and some of the other mixed races. The heathen, as you call them, are as corrupt as they can be. The Chinese, for instance, are almost on the verge of self-destruction. Unless the gospel reaches them they must perish as a nation altogether. Spirits and devils could not make them any worse than they are."

"How long has this impressibility existed, and what are the consequences resulting from the extreme sensitiveness of these races."

"The susceptibility to impressions from this world has been gradually advancing as the means of comfortable living and educational facilities have increased. The consequences to the earth life are in one respect very sad, but in another very hopeful and cheering. First, the sad and evil effects are that we, the powers of darkness, have been enabled to infuse into the German mind their system of rationalism and transcendental philosophy. Fine spun it is, and very plausible, but totally at variance with the spirit and temper of the gospel. Second, the hopeful and cheerful aspect consists in what we think will in time usher in the millenial dawn, which is evident from the fact that the world of earth life is progressing to a condition of sensitiveness that will eventually enable the angelic beings to communicate with the earth life even as we do now, the consequences of which will be untold good to the world. But that time is in the future, and you can compute the years that will intervene between the present and that time, perhaps as well as ourselves. We think, though, that it is not many centuries distant, but we reason only from analogy. The last century has witnessed great progress in the intellectual sensitiveness of the race. We think the conditions that enabled holy men of old to write or speak as they were moved by the Holy Spirit was a forced condition for special purposes, and not a general and natural condition. But your scriptures indicate what the signs of the times even now shadow forth that the period is not far distant when they shall all be taught of God; that is, through the instrumentalities appointed by him."

"But the present revelations from your world would not result in so much evil to this world, provided the sources whence they come were rightly understood."

"Most certainly they would not. The writing and speaking of the modern mediums appears so supernatural and marvelous, that many are ready to attribute the strange phenomna all to the power of God, and strangely believe that all that may emanate from spirits must be the all of truth. Many, in the simplicity of their faith, surrender reason, conscience, and virtue itself. The world will become wiser by and by, especially when they learn to know that these spirit manifestations are clearly prophesied of by the apostle Paul in these remarkable words: "And then shall that wicked be revealed, whom the Lord shall consume with the spirit of his mouth, and shall destroy with the brightness of his coming: *Even him* whose coming is after the working of Satan with all power and signs and lying wonders, and with all deceivableness of unrighteousness in them that perish; because they received not the love of the truth, that they might be saved. And for this cause God shall send them strong delusion, that they should believe a lie: That they might be damned who believed not the truth, but had pleasure in unrighteousness." This passage sets forth the signs of these times so clearly that all the righteous or pious can clearly understand.

Besides, there is much of good and truth in many of the communications. This fact will be a source of the greatest deception, as it was to you; but as you intimate if your world could rightly understand the source from whence all the lying wonders emanate the injury would not be so great. As it is, the elect, the chosen of God, they shall understand. It is not possible that they should be deceived because they are taught of God and are the temples of the Holy Spirit, which spirit will lead them into all truth."

"How does it happen that many are deceived in reference to their *relatives* communicating messages to them through the mediums?"

"It occurs just as it occurred to you. There are thousands of wicked spirits who are as familiar with your every act as you are familiar with the acts and thoughts of your own household. And they are capable of any amount of deception, having the power to control and shape the destiny of many little events in the earth life."

"Do not many wicked persons, upon their

arrival in your world, for a time think they are in heaven?"

"Perhaps they may, for a short time, but they soon learn that they are not in heaven, and from this fact: The spirits who have preceded them, and who are of like character with themselves, gather around them and by their profane words and lying speeches soon convince them where they are. Many of them rejoice for a time because they are not in a flame of fire, but they soon realize that they are in an element which gives them more suffering than fire could the body. Because there is an element which pervades the air, call it magnatism if you choose, so volatile and all-pervading that every thought which proceeds from a wicked mind is borne upon this element and comes in contact with other minds, rendering it impossible to escape the moral malaria which infests every part of the air. There is no possible method of avoiding these influences. Like a number of persons submerged in a filthy pool, all necessarily become alike filthy."

"Can you assign any reason why so few

are saved when the gospel is at every door?"

"The reason may be attributed to the fact that so many at the present time are impressible. The numberless theories of phylosophy which are entertained in this world are infused into the minds of the earth life. This leads many to be dissatisfied with the plain doctrines of the gospel. The world is running mad after wealth. Vast multitudes sell their souls for a mess of potage. How true are the words of your master, Jesus, 'How hardly shall they who are rich enter into the kingdom of heaven.' Not but that men may enjoy religion, pure and undefiled, and at the same time possess wealth. But the possession of wealth so stimulates every latent evil in the nature of men that it thus becomes a stumbling block over which they plunge into the dark world of woe. We positively know whereof we speak." * *

CHAPTER XIX.

I subjoin here another conversation with one of the invisibles that surrounded me:

"You have, as a man, a very faint idea of God, as the maker and controller of the universe."

"I should think that man in the earth life generally possessed as fair an opportunity of understanding the nature and attributes of God, as any invisible creature upon or above the earth?"

"By no means. You are in the shell of the earthly form, and what can a man in the form know of the creation of the universe? He has not fully opened his eyes upon God's works."

"Do we not know and understand as well, or better than spirits, in and above the earth, the nature, power, wisdom and glory of God? Do we not understand correctly God's creative power, by which he made the world?"

"You understand the nature and attributes of God better than some spirits, because you are more intelligent than some spirits, but there are some spirits within the earth's atmosphere, who know theoretically more than men in the earth life."

"You have admitted that all spirits within the region of the earth's atmosphere to be wicked, and by a law they are confined to that region, and can not therefore, learn by observation more than those in the earth life. The revelations in nature and the Bible, give the most exalted ideas of God, and we certainly have greater facilities for reading them both than have spirits."

"Very true, men in the form have the same objects to behold that we have, but we see with other eyes than they. Disembodied spirits look upon the spiritual uses of every thing, that is, they see the spirit essence which is the abstract quality of all substances. They see the uses of things as those in the earthly life can not see them; besides they are above and around, and have the power of locomotion, as none in the earthly form have, which enables them to observe critically, all parts of the earth. They also have all the accumulated knowledge of the past at their command—the wisdom of centuries. We have with us, in the air, as you call it, all the wise men of the past, who have not attained to eternal life through the gospel, therefore, we must know more than you. It is true, we generally reject the revelations of the Bible, because those revelations cut us off from all hope of future, endless life, and we instinctively reject them as a drowning man would reject the last hope of salvation, and that beyond his reach. We think the Bible contains many wonderful revelations which we know to be true; some, we think, we know to be false. For instance, the Bible says, God created evil; this is true. It says, also, that man is prone to evil as the spark to fly upwards; this is true. But when it condemns men to eternal suffering for being what God made them, we believe that to be false."

"We answer, that our most learned men must certainly possess better opportunities of searching for hidden truths contained in the book of nature, than the wicked inhabitants of the air, for the reason that they can operate upon crude matter in their investigations, but spirits of the air possess no such power. You can communicate to the earthly life by impression, and through writing mediums of the present day. But in this way you have not exhibited any superior knowledge, especially in the direction of the various sciences you have revealed comparatively nothing but what was before known. If spirits know more, surely it would be shown in their revelations. A man is known by the wisdom that proceeds from him. If you were truly wise, you would be good, and being good would lead you to impart that superior wisdom, which you say that you possess. God creates evils, such as tornadoes, earthquakes, and the like, but never created sin, for sin to be sin, must be the voluntary act of a free agent. But how do you know that God creates evils but from the Bible, which you believe to be untrue? That man is prone to evil is evident from observation, therefore you can speak definitely in regard to that subject. You admit that Jesus Christ was inspired by a wisdom infinitely above that yet revealed by the wicked intelligences of the air. Why not then receive him and his wisdom?"

"Very true, we have shown no great amount of wisdom in the direction of science, but we have revealed the fact that spirits inhabit the air of your earth, which, before spirits manifested themselves through mediums of modern days, remained a matter of very vague import. And then we have revealed the fact that the air contains the moral filth of the earth, and that all the morally pure ascend through the air to that world of light which we get glimpses of, but can not reach. We have revealed, also, that the Bible was written by man, inspired by Holy Spirits, rather than by wicked spirits, as at the present day. We admit that we are wicked, as are all in this department of God's creation, but still we think we know more than those in the earthly life. The reason of our being wicked, principally arises from the fact that we must receive all the moral filth of the earth, and by the time we get any portion of it reduced to order, there comes from earth multitudes more, and so on. You must not suppose the Devil's kingdom to be all disorder, for if it were so he would have no power over the earth life. Some who arrive in this world from the earth are anxious for the success of the gospel; this we soon eradicate from their minds. By their education in the earth life, they believe in its power to reform. They have also observed that those who came under its influence were rendered morally purer than by any other system of religion in the earth. They have also observed that all in the earth life who do not embrace the gospel (or rather whom the gospel does not embrace), never attain to the heavenly life. This earth-bias, as we call it, or predilection in favor of the gospel, must be entirely eradicated before they can be induced to embrace our phylosophies. The cause of your being so roughly handled was to ascertain if it were posssible for one of the elect to be deceived. Of this we are now satisfied. The elect of God are those who have experienced the implantation of that eternal life principle which is of God, and therefore we conclude they can not be deceived. The spirits who experimented at your expense were not all alike wicked. Some possessed earthly attachments, and therefore felt like showing some little mercy, but all were fully determined, if possible, to ascertain whether there was a higher life principle in the so-called children of God than in those who are not recognized as such. This we have proved to our satisfaction. Therefore, to obtain control of you was the first and most important point to be gained. We assumed to be the Almighty, not because we had lost all ideas of him, but that we might prove what we have indicated above. We assumed to be Jesus Christ, not because we have no idea of pure disinterested benevolence, but to carry out our plans. We led you from step to step until we made your God, as you believed to commit sin, and this you excused in him. Then we led you to believe that he had committed violence upon you, for this you freely forgave him, because, as you said, you loved him. Then we led you to believe that he was grossly profligate and wicked. Then you uttered these remarkable words: 'If God has done wrong he should repent, and ask forgiveness, that all others seeing his example, may be led to do likewise.' The Infinite God must have given you those words to shame us, if it were possible, for the unprovoked abuse we had heaped upon you. In order to farther test the fact that you were the child of God, and to farther impose upon your credulity, we said to you that God had been deposed from his throne, and that you could occupy it by our consent. This you would agree to do only because the good of the universe demanded it, and still you seemed to feel as kindly toward God as if he were your ruler. It has greatly surprised us that you have so steadily refused to acknowledge any relationship to any of the spirits around you, and for this reason, you could not believe your relations could be so wicked as to impose upon you, especially in such a shameful and unprovoked manner. This seemed to us the more remarkable because we continually infused into your mind the distinct impression that it was them and no others. But as it was against all reason, as you said, that all your relations should be wicked, and more wicked than in the earth life, you positively refused to admit the fact, and declared finally that that course was at an end. We admire you for this, because it showed a higher principle of action than is ordinarily to be found in the earth life. When pressed continually with the sssertion that your relatives had been indirectly the cause of all your troubles, you declared, " Well, if it is so, I freely and cheerfully forgive them, for if they are so morally weak as to injure one who has ever loved and cherished their memory, they are to be pitied rather than censured." You have demonstrated beyond all possibility of a doubt that you have a higher principle of something that is certainly divine, and not of earth, therefore we respect you, though we can never love you. We respect you in the same sense that the boy respects his hero in the tale he delights in. We shall make no more experiments with you, not because we could not control you, but because we have too much respect for you. We see that the world is running wild and rampant into the very ditch which has engulphed us. Therefore we have been provoked to reveal to you the fact that all the revelations through mediums to the effect that all men are progressing to a state of holiness and happiness is false, totally and absolutely false. We have as good an opportunity to know the facts connected with all the modern revelations as any spirits can know in this world, and we certainly know that they are not of God, but from spirits some of them guilty of greater abuses, if it were possible to inflict any greater, than we have inflicted upon you. And as you have persisted in the determination to expose the whole of the so-called revelations of spirits, we are determined to aid you; not because we fear God or regard man, but because you have commanded our admiration, and have demonstrated that there is a pure and undefiled religion in the world, though much counterfeited. In conclusion, we have no hesitation in saying that you are the purest minded man in the world, not from assertion, but from actual demonstration, for these experiments have not been confined to a few weeks, but have been contined through many months. You are entitled, we think, to be called a saint, from the definition we gave you, viz: 'A saint is one that would not do a wrong even if permitted.' We have also noticed that your prayers are directed to God your Father, and there seems to exist between you and him an intimate relation, a filial feeling, and we have doubts whether another could be found in the world, who possesses it in so eminent a degree."

CHAPTER XXI.

"You ask us how we know that modern spiritual demonstrations are not of God?" We answer that we know it from the following reasons: These spirit demonstrations are to my knowledge made by spirits who hate God, and have no fellowship with that which is good. You may know that they are all from this world by the fact that they universally reject the Bible as the word of God, denouncing it as a fable, and unworth of belief. This is characteristic of all who have been any considerable length of time in this world. All the good and pure love the Bible, and have always loved it since the world began. This ought to be conclusive evidence, at least to all the good and the pure, that the communications of modern spiritualism are not of God. We think these manifestations were designed simply to *demonstrate* the existence of the spirit world called in your Bible hades.

"You in the form have but little idea of the immense power for evil that this world exerts over the inhabitants of earth. It was also designed to prove the Bible true, not only in respect to its declarations with regard to heaven, but also in reference to hell. The revelations of the Bible were made, we believe, by angels and spirits appointed for that purpose directly by God, while the modern revelations are virtually from hell. This is true from the fact that all the revelations yet made by spirit manifestations have not so much gospel truth in them as has yet resulted in the regeneration of one soul in the sense that Jesus Christ taught regeneration. The revelations of these spirits are just what you might expect from beings who have not the love of God in them. They are characterized by any number of theories with regard to the origin of the world, and with regard also to the origin of the human race. They are philosophies brought here from your world, and consequently there are as many theories of world origin as have ever existed in the earth life from whence they sprung.

The heathen are all here in this world, and of course they have all the crude theories which they derived from their false ideas of religion. Consequently there can be nothing reliable revealed from this world, except the fact that all are wicked—not fools, for we retain all the wisdom we acquired in the earth life—but really wicked in heart, and deprived of that love of holiness which I now know springs only from a nature renewed by the power of God.

"From statements made in the Bible, and from indications in nature, some of us are of the opinion that spirits in this world are finally to perish, or as we think, to die a spiritual death as you die a physical death. For it is said in the scriptures: 'The soul that sinneth it shall die.' Others of us are of the opinion that we shall be burned up in the great day, when the elements being set on fire, shall dissolve the earth and all things that pertain to it. The atmosphere which constitutes our abode of course must perish with it; and hence we must perish for want of a place to exist.

"We are not in despair, because we have

much to gratify our senses, to engage our attention, and if we did not love evil and delight in it we could be made happy. We are not, as some of you suppose, roasting in fire; our punishment consists in being morally filthy, combined with all the misery which arises from a community made up of wicked, designing, and wholly selfish persons."

QUESTION.—"The Bible speaks of the prince of the power of the air, or in the air; what may we understand by this?"

SPIRIT.—"You will understand that the Prince' is the name of the ruling spirit of evil. There are many spirits in the air, who are rulers, just as Indian chiefs rule the tribes to which they belong. You have but little idea of the condition of society in this world. We have all that is vile and polluted of earth."

QUESTION.—"Is your society divided off into nations and communities, with the manners, customs, and habits, peculiar to themselves?"

ANSWER.—"Yes, each nation retains essentially its national peculiarities. Their moral habits and soul degradation remain the same. They ascend into the air a very transcript of earth. They are turned into hell, because that in the earth life they are already in a degree of hell. Sin produces suffering, and suffering in full measure is hell."

QUESTION.— "Are there any mediums, known as such, in heathen countries?"

ANSWER.—"Not to my knowledge,—they are not impressible; if they were, dreadful would be the wrongs inflicted upon them. You know something of the suffering which results from impressibility, even in this country, where society is a thousand times better than in that of the heathen."

QUESTION.—" Why are they not impressible?"

ANSWER.—"I can not tell you. I suppose it is owing to the fact of their thick skulls and want of nervous sensibility?"

QUESTION.—"Are not the wicked who leave our state of society more vicious and outrageously injurious than the wicked who reach you from the heathen world."

ANSWER.—" Generally they are, and it is, I

think, because of the law which runs thus, 'He that being often reproved hardeneth his neck?'"

QUESTION.—"Are there no good angels about the earth, and have they no power to correct the evil, the awful abuses of these wicked spirits?"

ANSWER.—"None at all that we know of. There may be thousands, but we can not see them. I think they have very little power to restrain the evil, even if they were present. But that they will have more power at some future period, power to impress and influence minds in the form, is evident from the fact that the whole race is becoming more and more susceptible to impressions?"

QUESTION.—"What do we understand by the Savior's expression, 'And there shall be weeping, wailing, and gnashing of teeth?'"

ANSWER.—"There is that which corresponds to it. There is no shedding of tears, but as an expression of overwhelming grief, it is true, prositively true. As soon as a soul realizes the fact of its being lost, fully and irrevocably lost to all good, to holiness and heaven, then such expressions of grief as proceed from the poor victim of sin is enough to make all hell shudder, if they were good. But as they are not, the wailing is only echoed back in derision upon the heads of the poor, grief-stricken victims and with fiendish contempt."

"Is not the declaration of scripture literaly true, which speaks of Jesus being tempted of the devil?"

"It is true, I think, in the same sense that you were tempted by the devil, or by devils. We think he was impressible like your self, and consequently exposed, as you are, to the power of the air. Some with us are of the opinion that he was led into the wilderness for purpose of developing his mediumistic powers. But this can not be true, for he, like Samuel the prophet, must have been very impressible from his youth, at least, before he was twelve years of age, as he then reasoned with the Doctors in the temple, and astonished them with his wisdom. Therefore we infer that he must have been impressible, and if so he was taught of God or taught by those appointed by him; and being impressible he must have been exposed at times to the powers of evil."

"Does your world possess any correct ideas of God and his attributes?"

"We have the same ideas of God that we possessed in the earth life, that is those of us who received a Christian education— nothing more and nothing less. The heathen have the ideas that they obtained while upon the earth. We think one reason that the gospel has made so little progress among them arises from the tenacity with which they hold on to their earth education. Believing it right, they surround those who are in danger of being converted to Christianity with all all the obstacles and difficulties that their united wisdom can invent."

CHAPTER XXII

"To what extent have the powers of the air dominion and rule over the children of men?"

"To reveal *all* the sources of power, which belongs to the rulers of darkness, so-called, would take a folio volume. But I will state a few of them:

First. The powers of evil have the means of creating electric currents around the most positive man upon the earth, and by so doing enable some one of our number to make a distinct impression upon his mind. This has often been carried to the extent of controlling some to commit acts of violence upon their fellowmen.

"Second. The powers of evil have the means at their command of concentrating currents of electricity along the telegraph wires so as to obtain the intelligence which is transmitted.

"Third. They also have the power to rule

with a rod of iron the poor victim of intemperance. This is done by coming in rapport with him, which implies spirit connection and while thus in rapport draw the electric influence of the ardent spirits into their own persons. This creates the same sensation in the spirit he realized in the earth life only, not with the same injury. This coming in rapport with the drunkard, in the manner above described, is very injurious to his mind—weakens the will power and renders him still more the victim of the bottle and of other vices.

"Fourth. They have power to use light substances, such as vapor, particles of dust, and the like, and by attracting these around them they are enabled at times to manifest themselves in human form. Thus they have frequently appeared to you and to others. But this form is no more the real spirit than the clothing is the real man. This power is often used to deceive the poor victims of mediumistic clairvoyance. They are the more deceived because they see the apparition with the naked eye, and suppose they therefore see a spirit.

"Fifth. They have the power to produce life-like images in the mind of impressible mediums. This is oten understood by them to be an actual sight of a real object. This leads to a great variety of delusions, and to such an extent are some imposed upon that they are carried away with the idea that they are prophets, and by these visions attempt to foretel future events. Those who are called leaders of spiritualism, and who know the fallacy of these impressions, allow the deceptions to go on and are therefore participators in the swindle. This stamps them with infamy.

" Sixth. They have the power of using the human body, with all its organs and faculties. This is done in the case of trance speakers and personating mediums. They enter the body by means of electrical and galvanic influences, and having entered they use the vocal organs just as they did those of their own bodies whilst in the earth life. Once in possession of the body, the spirit of the rightful occupant is stultified and compressed to less than half its original dimensions. This often injures the spirit thus treated to a very great extent, weakens its powers of reasoning, of memorizing, and of abstraction.

"Seventh. They have the power of collecting electricity and discharging it against a board, table or any other hard substance that is sonorous, producing concussions equal to the ticking of a watch or clock.

"Eighth. They also possess power to move ponderous objects, such as tables, chairs, &c. This is generally accomplished by the agency of scores and hundreds of the invisible workers, not always in the house, but hundreds of feet in the air above.

"Ninth. They possess the power of injuring those in the form by shocks of electricity concentrated upon them from the clouds. This has often been done, too, in a clear sky, and without any concussion or noise. The persons thus killed are supposed to have died by apoplexy or by an internal rupture.

"Tenth. They have the power of opening doors by raising latches, drawing bolts, and removing any other small obstacles or obstructions. This is effected by means of electric currents concentrated upon the object to be moved.

"Eleventh. They possess the power of controlling some men while in the act of writing, so as often to change or alter the form of a letter or figure, by so doing ruining them in their business. This takes place often in drawing up instruments of writing which convey the right of property from one man to another."

"Will you give me the reason why all mediums are not abused by the spirits who surround them?"

"The first, and to me the most obvious reason, is the fact that many mediums are too positive, that is, they have too much power over their own bodies. If the strong man keeps the house how can another enter and take possession? Another reason lies in the fact that the medium is in moral fellowship with the controlling spirit. In that case there can be no quarrel between them, but a perfect agreement. Any two individuals, however low or abandoned, can walk together, provided they occupy the same plane or moral level. In your case we found no moral sympathy. We found you on a plane so vastly above us that we could not feel the least fellowship for you. We endeavored at first to lead you to embrace our philosophies, but finding that that they would not go down—your mind being too firmly fixed in what might be called the divine philosophy of your religion—we therefore were obliged to abandon all ideas of converting you to our purposes. You would not walk our way and we could not walk your way, therefore we fell out and abused you without stint or mercy."

"Another reason why all mediums are not abused arises from the fact that the controlling spirit may have lately deceased, in which case their earth life sympathies lead them to be kind and very careful of the medium's comfort. and well-being. The most abandoned and wicked men at heart may have been so far under restraints which the usages of society impose that their real character is not apparent, either to themselves or others. So after death these society-imposed restraints continue often for a considerable length of time—in which case should they become the controlling spirit over any medium, they would not treat them in accordance with their interior nature but according to the society-imposed nature; hence you will understand more fully what is meant by earth life sympathies."

"Another reason why mediums are not always abused is because that many spirits entertain the idea that their philosophies will make a mighty movement in the world. We mean by a mighty movement, a grand commotion, a great excitement. Many in this world, like many in the earth life, have their pet theories, and they will compass heaven and earth to make their philosophies successful as a system of belief. When this is the case with a controlling spirit, and the medium is more than willing to surrender himself to carry out his plans, then there will be no abuse, but on the contrary great care is taken of him and every attention paid to his comfort."

"Have you no special system of philosophy

which engrosses your attention and heart sympathies?"

"We have none especially. We came to this world deeply imbued with a sympathy for the gospel system of truth, and we have not yet had occasion to change our minds or our moral predilections. You must think strange that we do not fall in with your plans of evangelizing the world. The reason lies in the fact that every thing, whether in the vegetable or animal world, unfolds according to its interior nature. For instance, the rose plant will, if it unfold at all, unfold so as to produce a rose; the thistle, with any amount of culture, will never unfold so as to produce a lilly blossom, nor will the lilly ever so unfold as to produce the thistle bloom. So you will understand from this exposition of the divine constitution in the nature of things, the reason that we can never have any moral sympathy with you. Our natures have unfolded according to their interior constitution, which makes us just what we are. We are as true to the unfoldments of our nature, as the thistle is to its nature, or you to your nature. Therefore do not blame us for being what we are.

"Your nature, and the natures of all true Christians, must unfold according to their inner life principle. One of your teachers has said, 'He that has this hope within him purifies himself even as God is pure.' Now, my reason teaches me that to become holy the man *must fall in love with holiness*. Society may impose upon man restraints, but when left perfectly to themselves they will act according to their ruling loves, their nature or their interior being. Hence, the world over the various Christian churches divide off into two general classes—the one embrace those who have the interior self purifying love principle and are therefore pious from principle, the other class might be called formalists—they adopt a form because society imposes it, not that they are really attached to one form more than another, only so there is no element of real piety within its inclosure. Their religion stricty defined is simply opposition to true religion—they are not in moral sympathy with the cross of Christ. You may think strange that I should know all this and yet possess a nature opposed to God, and having no moral sympathy with his children, but it will not be strange at all when you *learn to know* that what I have told you respecting the unfoldments of nature is the truth, and nothing but the truth, a truth which a man is not long in learning when he gets to this world. There are thousands in your world who have just as clear moral perceptions as I have shown you that I possess, who can point out clearly the difference between a true and false religion, but still have no moral sympathy with the true. Jesus, your Savior, gives as good a reason for this condition of mind and heart as I can think of, in these words: 'Ye will not come unto me that ye might have life.'

Thus kind reader, ends another commuication as remarkable as any which has preceeded it. I will simply remark that this document reveals more clearly than any which has preceded it, the real intention and designs the powers of darkness had in deceiving me. This was received since I began to write out the history of my sufferings and trials. Therefore it will be proper for me to say that it alludes to events of which I have not as yet spoken—events which I had intended to have covered with the mantle of oblivion. But as they are alluded to in this communication, I can not well avoid an explanation of the matter.

The part alluded to is in these words: "We led you from step to step, until we made your God, as you verily believed, to commit sin. This you excused in him. Then we led you to belive that he had committed violence upon you; this you cheerfully forgave in Him, and then you uttered these (to us) remarkable words: "If God has done me wrong, he should repent and ask forgivness, that all others seeing his example might be led to do likewise."

That I was at first led to believe that the voice, which announced itself as the Almighty was really and truly what it purported to be, I have already stated, and also the circumstances which conspired to bring it about. I will now say that at one time I was led to believe that God had not only *deceived me, but was even wicked*, and the direct cause of all my suffering. In reference to my stating that if God had sinned against me, he should ask my forgivness, I remember distinctly that I said it. I know, too, that I was as rational as at this moment. I was only deceived in the source of my information. It may seem to the reader that I need not have been deceived if I had resorted to all proper means of correcting my judgment. I have already stated some of the measures to which I resorted, that I might arrive at the facts in the case, but I will briefly recapitulate them: First, I carefully weighed the substance of the communications to see whether they confessed Jesus Christ, and they did confess him, fully. Second, the voice announced itself as the Almighty. I had no means of obtaining any facts to the contrary, therefore, as I owed every thing to my creator, I felt the most solemn obligations to do his will. Third, this voice corresponded in description to the voice which spoke to Elijah in the cave of Mt. Horeb, viz: "A still, small voice." It seemed not addressed to my ear, but, as I have said before, to my mind. Besides, the voice itself announced that in like manner it spoke to prophets and patriarchs of old. All this was done in the most solemn and awful manner. Think, reader, how you would have met these trying circumstances, and then pronounce your judgment accordingly.

I will call the reader's attention to these words of the spirit: "We impressed you continually that we were your relatives, and had been indirectly the cause of your sufferings." The power which these invisible beings of the air have to impress the minds of those in the form, must be immense. They have power not only to impress words and ideas, but their own feelings. My experience of this has been the cause of much of the suffering which I was compelled to endure. Some times, for hours, I would be impressed with a feeling of irritability or anger, without the slightest cause. This strange impression would usually be preceded by these words: "I am mad at you." At other times an undefined impression would overshadow me, compelling the belief that a deceased brother or sister was near and around me, and causing the severe developing pains they were inflicting upon me. This feeling or impression I could not shake off or remove. It would last at times for days, unless I could prevail upon them to remove the impression. This they could do in an instant, and again cause it as quickly to return. They could also imitate the manner of speech peculiar to my relatives, and acquaintances, and so exactly did they give the particular intonation and inflections of voice, that I would have been compelled to believe the imitation to be real had they not personified the voices of some whom I knew to be living. Upon one occasion that occurs particularly to my mind the voice, style of address, intonation, &c., were so exactly personified, that for the moment I felt positive that the gentleman and lady represented had deceased, and that their disembodied spirits were before me. But when I knew by the evidences of my physical senses that it was not the case, I was then convinced that they were presenting assumed characters. I could but be astonished that so exact a personation could be given, and especially of persons still living in the form.

This personation of character at times was carried to great lengths. The persons or characters introduced, were generally distinguished men of former years; men who had made their mark in the world of letters, theology, metaphysics, &c. These characters were usually introduced in groups of two, three, or four, each one expressing his opinions as the conversation turned upon this or that part of the subject under discussion. These personations often continued twenty or thirty minutes. At times, I consented to take part in the debate myself, asking questions, giving answers, and expressing my opinions. When Dr. Chalmers, Dr. Brown, or Dr. Abercromby were introduced, the conversation was made to turn upon theological or metaphysical subjects, such as the origin of sin, human accountability, the sovereignty of God; or again, upon the conscience, the reasoning faculties, or the varieties and phases of human intellect. If political men were introduced, the conversation turned upon the condition of the country, the war, slavery, &c. After a time, characters were introduced who, I felt sure, were still living when this took place. I remarked with surprise, saying that such and such persons were still in the earth life. "Very true," said they, "but we have the power of bringing you in rapport or electrical connection with them, which answers the same purpose as if they were present." This I felt sure was most glaringly false, and that all the characters were assumed for the occasion.

Thus by degrees I became acquainted with their habits of deception. For a time, they attempted to make me believe that repentance and Godly sorrow for sin, were as common in their world as in our own. This they did by bringing characters to me, who seemed to realize a burden. The character, introduced would begin in the most penitent manner to speak of their burden of sin, of their deep sense of guilt, of unworthiness before God, and of earnest desire to receive *God's pardon-*

ing favor. Then in the most beseeching manner, the character would request me to entreat God in their behalf. This penitent demeanor would continue some times for days, then as if suddenly experiencing pardon or conscious forgivness from God, they would exclaim, "Glory, Glory to the Highest." Then the character in a few moments would say, " I am going to leave you; God has called me up higher," Then with a shout of glory, again, the character would depart, shouting faintly, and still more faintly, until the sound seemed to die away in the distance and cease altogether. For a time, all would be quiet. Then another character would be introduced, inquiring for the person who had been mourning, and when informed that they had professed conversion and had gone up, as they said, higher, the new character would say, " Well, I have been expecting this for some time. I was quite certain we should loose her. She has seemed of late to wear a sad and serious countenance. She had but little to say, and never joined in our revelry. I was sure God was preparing her to be taken out of this place." Thus would the second character continue to talk about the first character for hours. Then all of a sudden the first character would return apparently in great glee. Both then would join in a general run on me for being so easily deceived, assuring me that I ought to have known that there could be no repentance in the world of lost spirits.

Thus from day to day they would demonstrate to my mind (not intentionally I think) that they were beings totally lost to all that is good, capable of blasphemies of the most horrid character, of deceptions in every possible form. Paul, in the 2d of Romans, draws a very graphic picture of their characters in these words: "They are filled with all unrighteousness, fornication, wickedness, murder, deceit, haters of God, dispitiful inventions of evil things, without natural affection, implacable, unmerciful.'"

My more than four years' experience of intercourse with these invisible beings has firmly fixed me in the following convictions:

First. They are intelligent creatures, retaining in the disembodied state all the vigor of mind that they possessed in the earth life.

Second. That they are, as I have stated above, filled with all unrighteousness and malicious wickedness.

Third. That most of them at least once occupied a human body. This I infer from the many cant expressions and phrases that I have heard them utter. Their positive declaration to the effect that they had once occupied the human form amounted to nothing towards convicing me of the fact, but their seeming familiarity with earth life, their many apparently accidental allusions to it, and their great variety of earth life phrases, I think gives some weight to the conclusion.

Fourth. That they are possessed of knowledge, bearing upon every subject within the range of human thought.

Fifth. That they have the ability to impress their thoughts upon the minds of those who are in the earth life, and consequently may properly be called princes of the power of the air, who work in the children of disobedience.

Sixth. That they have the ability to surround some in the form with electrical influences sufficient to produce psychological impressions, or mental pictures of every form, color, and devices known or imagined by man. Also that they are able to collect particles of dust, vapor, and other substances sufficient to make every form or shape in the heavens above, or in the earth beneath. And also have the ability to give the object presented any shade of color they may will to give it.

Seventh. That they have the power, not only of impressing, or infusing the thought or idea into the minds of men, but they have the power to impress some with the word, and even the articulate sounds of the voice. How this is done I am not prepared to say, but that it is a fact my more than four years' experience has tested under a great variety of circumstances. I am also convinced that they can produce impressions of their feelings upon some, as I have many times experienced to my great discomfort.

Eighth. They have the ability to inflict upon some in the form exceedingly sharp, electrical pains in any part of the body, and of continuing them for any length of time. These pains may be increased to the extent of starting the perspiration from every pore of the body.

Ninth. That these beings have no means of knowing what their final end will be except from the Bible, and they hate that book as all wicked men do, because it unequivocally condemns them.

Tenth. That all the theories of the origin of the earth, of the origin of man, and of the endless progression of all men to a state of holiness which have emanated from these beings, has no more authority than any theory emanating from earth life.

Eleventh. That by these manifestations from the other world, God designs to teach man that his word is true. That there is a hell, not of fire, but a hell more awful than is possible for the human mind to conceive—a hell resulting from the confining within certain limits, all the wicked from earth life and these limits being the earth's atmosphere, verifies the truth of God's word. "That the wicked shall shall be turned into Hell, and all the nations that forget God."

Twelfth. That salvation results from the operation of the Holy Spirit upon the nature of man, giving it a holy tendency, the results of which creates in man a desire for moral purity or a hungering and thirsting after righteousness. That the eternal life principle, or the love of holiness in the heart, is an active principle, and like leaven hid in the meal will work and permeate until all is leavened. That when the righteous die, they ascend to heaven, which is a place beyond the earth's atmosphere.

Thirteenth. That the revelations of these powers of the air have never recommended the gospel plan of saving men from sin, but have in every instance discarded it in the bitterest terms. They also reject the Bible, and substitute for it the endless and contradictory philosophies which they have given to the world through mediums. Though many communications from these beings are characterized by *much that is good*, and by professions of love and respect for Jesus Christ, yet they are only the gilded bait to introduce their philosophies, or as Paul characterized them: "doctrines of devils and demons."

Fourteenth. I am fully convinced that the beings who have been surrounding me are more wicked than any person I have any knowledge of in the earth life. More wicked because they commenced and have continued upon me at intervals for more than four years, a species of mental and physical torture without parallel in the history of the world. When wearied, worn and exhausted by continued suffering, still they manifested no sympathy whatever. In addition to the physical pains inflicted, as they said, to develope my mediumistic powers, they used every cutting, sarcastic epithet that mind could conceive of to irritate, tease and abuse me. By day it was every thing by turns, at night, visions, specters, ghosts and demons.

Fifteenth. I think there is evidence that wicked spirits have greater power over the earth life than those denominated holy spirits, or the spirits of just men made perfect. Their magnetism, or the electrical influence which they are capable of exerting over the minds and bodies of men, is, so to speak, rougher and harsher than the magnetism of the good and pure who come to us from heaven. And I think we may infer from this fact the reason that all the communications from the spirits of the present day are from a a class of beings denominated wicked spirits or demons. The people of those nations who rank as the most refined and sensitive are those among whom these wicked spirits have manifested their power. This may, in process of time, be extended to all the inhabitants of the world. We may infer, also, that in the far off future, there will arrive a period when the present enlightened nations will have arrived at that degree of sensitiveness that holy spirits as well as angels will be able to influence the earth life just as these wicked spirits do at the present time; that this will truly be the millenial dawn, and the earth life will then experience the fulfillment of the prophecy: "that they all shall be taught of God," or by pure and holy beings, who are appointed for that purpose by the will of God.

This period will be known as the millenial Sabbath of the earth.

Sixteenth. From long continued intercourse with these beings, I think I am warranted in the conclusion that they possess all the intellectual faculties in the same degree of activity as possessed by any in the earth life. All that they seem to lack in regard to the element of human character is reverence for God, personal holiness, kindness and love. They possess the power of detecting the moral qualities in actions in a very eminent degree, but possessing it, still they lack personally every moral quality. As an evidence of their mental accumen and capacities of reasoning, I subjoin the following expression of opinion on a question which I propounded to one of these invisibles, and a criticism of that opinion by a fellow spirit I think of about the same intellectual power.

CHAPTER XXIII.

"Are evil spirits capable of exerting a greater power over those in the form than those who may be termed good and holy spirits, or the 'spirits of just men made perfect?'"

"I answer, yes, and no. Evil is ever powerful, only less powerful than truth. The truth is powerful and must prevail. It ever has, and it ever must. But we can never know the all of anything. To know the all of truth, is to know the all of God. Power can not come from two sources. Therefore the mighty power of God, will and must prevail; and by that almighty power, all things shall work together for good, even your good."

Then follows the criticism. "The spirit who wrote the answer to the above question certainly intended to *deceive*, or give an *evasive* answer. It begins by asserting that it is both true and false; that evil spirits have greater power over those in the form than those that are holy, or morally pure. The spirit that gave that answer must know that evil spirits have vastly more power over gross matter, or over the minds and bodies of those in the form, than those who are termed holy. Here they are always used in developing the mediumistic powers of those known as mediums of the present day. It is also plain to be seen that the explanation in reference to the relative powers of good and evil spirits, over those in the form, is very evasive. It is deceptive and subtle in its reasonings, because it says that all power is of God—the evil power is a part of the all power,—therefore *it* is of God. The fallacy of the declaration lies in the second part of the proposition, that evil power is a part of the all power. It is not so, either directly or indirectly, because evil power is a perversion of legitimate power and God is not the author of perverted power, therefore he is not the author of evil. But it may be said that God permits all perverted power, which is equivalent to ordaining it. This declaration can not be true, and for the following reasons: God permits evil that he may learn his creatures lessons, which they will not learn by direct command. He says rejoice when ye are tempted, or tried, for the trial of your faith is of more value to you than gold or silver. To assert that evil actions are of God is to make God the author of sin. God makes creatues holy and they voluntarily make themselves *unholy*. It must be evident that God never ordained the action of a rational, accountable creature, because he has first ordained his freedom, and to ordain the act of a free agent would be to unordain his freedom, which is equivalent to God acting against himself. Therefore to assert that God is the author of all power, including the power exerted by evil doers, is false and can not be sustained. But the direct answer to the question, ' Have evil spirits a greater power over those in the flesh, to control the minds and bodies of men, than those called good spirits?' This question admits of two answers. Evil spirits can exert considerable power over those in the flesh, but it is limited and for the reason that wicked spirits have no power to control the minds of men only by impression. If men reject the impression, they can do no harm. But in reference to the power of evil spirits over the bodies of men, I would say that evil spirits have power over the bodies of men, so far as mesmerizing, or psychologising then is concerned. This is done in precisely the same way that it is done in the form. Therefore possession of devils is true when ever the so called devil or evil spirit has gained over his subject mesmeric control. This was more common, I think, in the days of Jesus Christ than it is at the present time, from the fact that men are now physically more sensitive and mentally more positive; while evil spirits are more negative, which separates and gives to evil spirits less control.

"With regard to the influence of good, or holy spirits, over the earth life, we judge from the fact that they are never here, we think, to exert much influence except during the continuance of what in the earth is called a revival of religion. Then they give evidence of their presence, and are enabled to exert considerable power over those in the form, not physically, but mentally, that is, by impression. That they ever develop mediums we know is not the case, for all mediums are developed only by those who are called in the earth life rough, positive, determined minds of the dare-devil stamp. That the time may come in the far off future, when angels so-called and spirits of just men made perfect, may be able to exert a direct and positive influence over the minds and bodies of men, I have not the least doubt. It must be evident that the human race, is fast approaching to that degree of sensitiveness, or impressibility that will admit of it. How long a time will elapse before that period shall have arrived, we are unable to determin. Your scriptures, I think, indicate the time as being between one or two hundred years' distant."

As another evidence of the intellectual capacity of these powers of the air, I subjoin an answer to the following question, " As nothing was created in vain, what purpose do you think the wicked subserve in the endless chain of sentient existences, beginning with the worm that crawls, and ending with the exalted seraph before the throne?"

"That is a question difficult to answer. We know of no use that they can subserve, except it be the refinement of matter. It must be apparent that the earth at one period of its preparation for the abode and temporary residence of man was without vegetation, so there must have been a time when vegitation began to creep over the earth. There was also a time, when no sentient being existed upon the earth; so there must have been a time when sentient existence began to manifest itself in the myriad forms of reptiles, insects, birds and beasts. The thought must have occurred to every reflecting mind, and the question propounded either to self or others, why has the creator brought these myriad forms of matter into existence? If this inquiry can be rationally solved, it may prove the key to unlock the hidden mysteries which surrounds the question you have propounded. We say then, without hesitation, that they were created for the purpose of refining matter. The *air, electricity*, water and fire, were engaged in the same work ages before the existence of vegetation, and were as essential to the production of every form of vegetation as that vegetation is essential to the production and sustenance of sentient or animal existence. If the elements had not continued the great work of preparing soils upon the earth, then vegetation would never have been produced; if vegetation had never existed upon the earth, animal life could never have been sustained; and if animal life in all its myriad forms had not still farther refined matter it must be evident that man, the ultimate and crowning glory of earth, could not have been sustained. As your Bible informs us that the angel reapers are not to harvest and gather into the garner of heaven all of the human race, therefore I think we may infer that the tares even if burned will subserve the great end of refining matter. Spirits subsist and exist upon the refined essence of matter which pervades the atmosphere, just as the finny tribes exist in water, and live not upon the water but upon the air, and finer elements existing in the water. So we, though existing in the air, live or subsist by virtue of the finer and more etherial elements which are held by that atmosphere. I think then we are warranted in the inference (reasoning from analogy) that angels and redeemed spirits exist, as we are informed, in the fine, soft, silvery sea which streaches out in every direction from the earth's atmosphere into the limitless ocean of happy being beyond. I say I think we are warranted in the conclusion that they exist, and subsist by virtue of the infinitly refined matter which is thrown off from this, and all other worlds into this soft ethereal expanse, and if there should prove to be a supercelestial heaven as we have heard of, made up of the still farther refined particles of matter thrown off from the myriad universes of worlds, which stretch off into space farther than thought of man has ever reached, even as far we might suppose as the utmost limit of the creator's thought—I say then if there should prove to be a supercelestial heaven, a heaven of heavens, made up of infinitessimal particles thrown off from the lower universes of worlds, there we might suppose is situated that beautiful city, the new Jerusalem, where they have no need of the sun nor of the moon to shine in it, for the glory of God did lighten it and the Lamb is the light thereof and the nations of them which are saved shall walk in the light.

"Thus you have my idea of the end and purpose that all the wicked and unsaved of earth subserve. You might think the knowledge we have of heaven and its glories, and the fact that we are lost to all that is good and pure and holy would be the source of our greatest misery, but it is not so to me at the present time. When first I emerged from the body into this strange abode, this receptacle of the moral filth of earth life, I was overwhelmed first with grief, then with rage, then I sunk down in despair. I cursed my existence, the day of my birth, and then my maker. Oh! thought I, to spend an eternity with fiends, with demons, and spirits damned. This caused in me grief, overwhelming grief. I looked up and around me, the same heavens were indeed above my head, the same earth beneath my feet, but all else how changed! Instead of the heavenly glory into which I had flattered myself I should emerge, I was surrounded by the same brick walls, the same old shabby out-houses, the same front yard, trees, and garden, and orchard. Instead of being surrounded by angel bands whom I had expected as an escort to convey my spirit to the home of the blessed, I was surrounded by horrid looking beings in human form; some of them I recognized as being persons whom I hrd know in the earth life, but greatly altered in appearance. Some of them came near me, and laughed and jeered, and mocked with fiendish cruelty. Others provokingly said you have arrived in hell at last; but cheer up, you are not in a hell of fire and brimstone. Others said if it is not fire and brimstone it is an element just as hard to endure. I attended my own funeral, heard the singing and the prayers, and the sermon just as in the earth life. I could discover no difference, save that I could hear more distinctly, and could apprehend more readily the ideas of the speaker. My mind seemed more keenly alive to all my surroundings. But alas, I was not good. The same impurities of mind, the same feelings of anger, malice, and hatred which had ruled me in the earth life, I found to be a component part of my being. In fact, I was essentially the same in every thing, save a corporeal body. I found that my locomotion was the same in every respect, only I seemed to glide along rather than move by walking or running. I very soon found that I could move or glide through the air much faster than I could pass over the ground in the earth life by running. I found that I could visit every place in the earth life towards which my inclination led me. I was several times to church, but discovered that I *possesed* the same *antipathy* to *gospel truth* which *caused* me to *shun* it in the *earth life*. In fact I found myself the very counterpart of what I was in my earth life, having the same antipathy to that which is good, pure, and of God. The carnal or unrenewed nature, I found to be as perfectly represented in my new life, as in the earth life. I felt a restless uneasy desire to do something, and that something that I wished to do was in opposition to the laws of my moral well-being. I was invited to attend the theater, and did so. There I found elements suited to my moral being. Here was to be found spirits who in the earth life were managers and prime actors, amidst the theater going element of earth life. I saw them surround their favorite actors, imparting a kind of inspirational ectasy, which enabled the actor to even astonish himself, and draw from the crowded galleries deafening applause. I saw them walk the stage and perform in unison with the actors of earth life, the same old dramas in which they took so much delight, when clothed with corporality, so that the spirit actors often outnumbered the earth life actors, and the spirit audience many times outnumbering the audience which still remained in the flesh. All these seemed to me at first very strange, but I soon became used to them, and they lost their charm. I then attended operas, dances, concerts, circuses, and every place where amusement and excitement were the order of the day. In these I experienced a measure of delight, but was far from being happy, because I was not holy.

"I learned enough of earth life, especially of earthly Christian' life, to know that excitement was not the element which entered into and formed the basis of true happiness. The highest exercise of my judgment gives me the clearest convictions that without holiness or moral purity there could not be realized true felicity and soul rest. I felt that I was cut off from God, and like a wandering star, there remained for me only the blackness of chaotica darkness forever. I felt that I was an outcast from all that was pure, holy and of good report, and then the overwhelming thought that I must sink lower and lower into the maelstrom of moral pollution, even a bottomless pit. I felt conscious that an irresistible current was drawing me nearer to that awful gulf into which the wicked eventually plunge, even the gulf of total and absolute alienation from all that is good and pure, loving and lovely. I felt that

I was becoming a perfect moral wreck; for I was being beaten upon by the wave, of moral pollution, without the least element of power to resist their force.

"The atmosphere of the earth I found to be the utmost limit of my range of locomotion; what existed above or beyond that boundary I could have but the faintest conception. Thus I found that I was a prisoner sold under sin, and so far as I knew, to be kept under these chains of spiritual darkness, unto the judgment of the great day, when I would be banished still farther from the presence of God and the glory of his power. If the recollection of the past could have been blotted out, and all remembrances of slighted opportunities of becoming good and consequently of becoming happy, could have been erased from the memory, I felt that I should then be comparatively happy. But memory, perfect memory, became the source of my greatest torment."

CHAPTER XXV.

"As I was saying that I visited every place where the votaries of pleasure and excitement were assembled, not even missing the haunts of the strange women or the haunts of the other more degrading vices, but found after a time that they all failed to interest or contribute toward that excitement which my nature constantly craved, I then commenced to travel. I visited, in company with others, every place of note recorded in history or mentioned by travelers. I noticed the manners, habits, and customs of the various races of the earth life. I contrasted their moral, intellectual, and physical condition with those of my own race. This only rolled back upon my mind the reflection that only those nations blessed by the presence of the gospel had any vestage of real happiness and soul liberty. How true the scripture seemed to me, viz: 'The wicked shall be turned into hell, and all the nations that forget God,' and the more so because I found that among the heathen nations as well as among those denominated Christians, that the death change resulted in no moral change; but that as in my own case they were all turned into hell, because that in the earth life they were in a measure of hell; and consequently must as certainly come to this world of woe and wretchedness, as that their hearts are alienated from all good. Justice or no justice, they all come into the same dreadful condition of moral being that those do who slighted the gospel, with but this exception, they seem to be more contented with their lot. Having never known any higher life, their minds are not filled with poignant recollections, as in my own case of grace, and offers of mercy abused. I found that the greatest proportion of the heathen believed in and craved most earnestly as their highest and ultimate good, *annihilation.* Whether they will ever attain their hearts' desire I know not. But the expression, 'The soul that sinneth it shall die,' would seem to indicate this as a fact, were it not that death in your scriptures is used generally in the sense of separation from all good, rather than total extinction of being. Hence my mind, not theirs, is harrowed up with the certainty that myself with all the wicked beings shall exist for subordinate purposes in God's universe; perhaps for the refinement of matter alone, and perhaps for other purposes.

"It is written, 'some vessels are made to honor, and some to dishonor.' The potter hath power over the clay, to use it as seemeth him good. I found also that the children of heathen nations, as well as the children of all Christian nations, are not found with us in the earth's atmosphere. They all, as we suppose, pass away, as do all true Christians, from the earth life, to the paradise of God above, and beyond the abode of the wicked, which, as I have said before, I found to be limited to the earth's atmosphere.

"After ranging around over the earth until my curiosity was completely satisfied, I returned to my native land, to try over again to extract some new delight, if possible, from the old round of sensual gratifications. But

I went out, as the scriptures say of the man out of whom went the unclean spirits, 'seeking rest in dry places, but finding none.' I began then to feel in full measure the utter wretchedness of my condition as a lost sinner. Here I was with every attribute of a human being but holiness or moral purity. I felt that God was even good to allow me to remain where I am. If he had taken me to heaven, with the nature I now have, I felt that heaven itself would be the severest part of hell.

"At length, in my rage for something new, I went to a revival meeting, thinking that something would turn up in the progress of events there to create an excitement. I seemed to crave excitement, as a fevered patient craves the cooling draught. I was not long a witness of these scenes, until I found that I could infuse into the minds of those engaged in the holy work, blasphemous thoughts, infidel notions, obscene imaginings, and the like. In this I found myself, with many others, taking real delight. We found that we could not only impress some with unhallowed thoughts, but could even excite their passions, inflaming them with lust and unlawful desire. Beside these methods of influencing the minds and passions of men, we learned that by forming a circle, made up of scores of evil spirits, around some of the most impressible, we could influence them to say many absurd things; and this proved to us a source of great delight. Sometimes we would bring our combined influences to bear upon some ungodly and impenitent man, stirring him to get up an opposition to the work of grace going on, and leading him to act as if he was possessed by the devil himself, as he was, in truth, by many of them. Thus our evil work reminded me of that often-quoted text, 'the powers of the air working, not only in the children of disobedience, but in many of those called saints; so that it was literally true, that the saints of God wrestled or contended not only with flesh and blood (that is their own corrupt natures), but with spiritual wickedness in high places.

"I often felt that it was too bad to thus torment those who were trying to do right, but others did it, and so I continued to do it, until I found it afforded me great delight. Thus I became conscious that I had sunk to the lowest depths of moral degradation. I even lost all sympathy for my pious relatives in the flesh; and found that I could torment them, and disturb their sacred moments as readily as those who were strangers to me in the earth life. About this time, some spirits made the discovery of a process of concentrating electricity, and discharging it against some sonorous substances, such as a board, table, or stand, thus enabling us to convey by the number of raps intelligence to those in the earth life. This operation awakened in us much delight, because by it we were enabled to astonish those in the form, and practice all kinds of deception and impositions. The knowledge of this process of controlling electricity spread rapidly all over the spirit world, until the most illiterate or vilest of spirits understood it as well as the most intelligent. This discovery was soon succeeded by others of a more startling nature, such as controlling the hand to write without the use of the medium's mind; controlling the vocal organs of others to speak and utter sentiments entirely in opposition to those held by them; and last of all, controlling the minds those in the form, so as to produce mental pictures or images of any person, place or thing which might suggest itself to the spirit operating. This last, and most masterly power, we have used to make multitudes believe that they could see at a distance without the use of their natural vision, but which as you well understand now by experience, is only produced by creating the picture of the person, place or thing in the mind of the subject, so that he is able to describe the object as perfectly as he could by the use of his natural eyes. This has given us almost the power to work miracles, and led many intelligent persons in the earth life to believe that it was all from God, through the agency and ministrations of angels, and spirits of just men made perfect; and this belief was the more confirmed in the minds of many because we could write, or speak out through mediums, many pure and exalted ideas; ideas that you now know came from the vilest of the vile, even from ministers of the gospel, corrupt men, who in the earth life, sold their souls, like Esau, for a mess of potage. These we employed to gild the bait so that even many worthy Christians have been beguiled and ensnared by these deceptions into a belief that these spiritual manifestations were all of God, when indeed they proceeded only from the powers of darkness.

"Thus matters stood when we discovered that you possessed mediumistic powers of a high order. We therefore resolved to put your piety to the severest tests of which we were capable. We had much doubt of their being a pure and undefiled religion in the world at all. We commenced upon you, at first by partially mesmerizing you, and continued to repeat this until we gained over you almost entire control. Of the deceptions we practiced upon you, it is not necessary for us to mention. It is sufficient for me to say that we all became satisfied of your piety, and all of us agree that there is a pure and undefiled religion in the world, and that there can be no going to heaven without it. I have been around you more or less for over four years, and can testify that you have not discovered the least evidence of moral purity about me or any who have been associated with me. I would say, in conclusion, that I have written out this narrative of adventures in the spirit world, not from any desire to give warning to the wicked, but that you may know that my history, with a few variations, is the history of every wicked person, who dies without the eternal life principle implanted in them by God himself, thus making them especially his own children, by a deep inward nature averse to evil, and in love with purity and godliness."

That the reader may be fortified at every point, and never be drawn into the belief that any communication from the spirit world can in any sense be from God (though it may breathe the very spirit of heaven itself, and be characterized by lofty sentiment, and the most elegant phraseology, and classic purity of style), let him remember that if such are given through yourself as a medium, or through any other medium, it will only be but the prelude or introduction to something monstrous and absurd. All my experiences of these beings who surround us in the air, sum up in this distinct conclusion: that they delight in evil as their chief good, and especially that branch of evil called deception. If any one thing pleases them more than any other, it is to make those in the earth life believe the most monstrous and absurd theories about spirit existence, their habits, manners and customs; the origin of the human race, and numberless other theories with reference to the birth, parentage and authority of Jesus, our Savior, and our relation to the gospel. I would exhort the reader, as did the apostle Paul, in these words: "Though we, or an angel from heaven preach any other gospel unto you, let him be accursed."

The most subtle method which these powers of the air use to induce belief in their monstrous absurdities, consists in making friendly allusions to Jesus Christ and his gospel, and in speaking very highly of some of its doctrines, and perhaps give a grand dissertation upon one of them, and in the mean time weave into the frame work of this dissertation, a subtle philosophy which would undermine the consistency of the whole, and render it perfectly absurd. A characteristic of this style of deception is to be found in the communication which I subjoin. It was given to me before I became acquainted especially with the method of deception, and of course I gave it at the time some credence. It is on the atonement of Christ, and begins as follows:

CHAPTER XXIV.

"Our ideas of the atonement are far below the angels, and their ideas of it are very far below the ideas which God entertains of it. If you would know what we think of the atonement, first know then that the human race is a mighty harp—touch but one string suddenly, and all the others vibrate in unison; also it must appear to you that the finer the chord the nicer it responds to that touch.

"When Jesus groaned on calvary the pain of his body and soul was shared in by every creature beneath God's infinite heaven, and his agony thrill still goes on sweeping through the countless worlds, and will continue its onward course until mankind shall go their way and sin no more. No human body is in health so long as a single atom of disease lurks beneath the granule of a bone, or between the cells of the most unimportant nerve of the viscera. Neither can society be calm, or the race happy either in time or eternity, so long as one unholy man or woman lives to mar the harmony, or become a discordant note in the great sympathia. For this reason the dwellers of the spirit land are impelled to action in behalf of those on earth by the first and greatest law of the universe —self-preservation. For in making man abjure his errors and turn toward the right, there is laid the foundation whereon to build the great temple of purity, perfection and bliss.

"The discovery of the great principle of unity; the acquisition of the positive knowledge, that every sensation, painful or pleasant, experienced by any even the most destitute, low, and degraded of the species, is shared *in by all the rest*, and from finding that the most sensitive, and finest nerved persons are also the most unhappy, we are led to infer the existance of a great vicarious law whose elements are sympathy, compensation, and distribution. True, some may and even do pass through the whole of this life and apparently escape the suffering, but not forever. God says substantially, 'bear ye one another's burden and so fulfill the law of Christ,' and borne they must be. Sensitives bear the greatest proportion of earth misery, and their fate at first sight seems to be a hard one—a life full of groans, tears and sorrows. Yet the law of compensation is operative through all stages, phases, and planes of being, be they mortal or immortal.

> 'He who the weariest paths has trod,
> Shall nearest stand to the throne of God."

"There are seasons when men and women of a certain temperament, and without any apparent cause, are plunged at once into overwhelming misery and woe; so overpowering as to be but faintly imagined by those of a different mould of organization. Is that often paled cheek, delusion? those days of almost bloody sweat a phantom? Will the physician's hypothesis of a deceased liver lungs, or heart, account for it? I think not; there is a deeper cause, a higher power in operation. It comes from suffering man; not a member of our race, can be overwhelmed with suffering without these sensitive ones feeling it. The wave of pain goes out, and as the needle is attracted toward the pole, so it is attracted towards those whose position on the plane of the great sympathetic nerve of the universe fits to receive it. Some one else receives it in turn, but in a less intense degree, until at last a faint and tinny wave reaches the throne of God.

"'Eloi, Eloi, Lama, Sabacthana,' groaned the dying Savior, and the throes of his agony went pealing through the universe, until the universal heart, even the majestic Prince and heavenly seraphim, quailed with agony. So still, as of old, there is operating the same great vicarious law. When the suffering soul turns meekly to God, relief comes; but not an instant before. This great law that relief is realized by the soul only in meek submission to the will of God, was known in very ancient times, even as far back as the days of Enoch and Abel. Do you see you sorely tempted man borne down by sorrow's wave? Help him from his gulf of woe, remind him that his deep distress has an answering cry. Whenever a sinful soul cries, 'Oh! my God' the answer comes, 'child here am I,' and in every, 'Oh ! my Father,' there slumbers deep, 'Here, my child.'

"I do not say that all sensitives are the mystic sympathants of those who suffer, for such is not the case; much suffering comes to them from other sources. But that a great deal of mental agony does come from the source alluded to is known to be true, not only through the ranks of angel bands, but by many on the earth. Much of the suffering extant in the world to-day, can be traced to the extraordinary sensitiveness now charracterizing such vast numbers of people, of which there are two kinds viz : The *natural* and the *hot house.* The former results from refinement, education, and right views entertained towards God, towards man, and towards the physical universe. The latter owes its origin to a general overworking of the brain and total neglect of the muscular system, also to improper diet ; heedlessness in clothing, as to color, texture, and amount ; disregard of heat, light, and personal magnetic influences. By reason of this treatment, of the physical and intellectual natures, the funds in the bank of life are exausted, at the very time when they ought to be the most plentiful.

"Among the subjects of attention at the present day, and really by far the most exciting is modern spiritualism. This subject, by reason of its intense and all absording revelations and wonders, attracts human attention to the exclusion of all others, causing people to exchange common sense, for phylosophies, not so useful; often inducing a sort of intellectual fever by lifting people above the earth life realities, and placing their heads in the clouds while their feet are upon the ground, thus causing men to be forgetful of their bodies, by keeping the spirit continually before the mind. When man cuts loose from God, Christ and the Bible, and launches into the Dead Sea of any ism, the result is sure to be destruction and soul ruin. Remember that spiritualism without God, Christ, and his word will unfit you both to live, and to die. There are thousands of people who should flee from spiritualism as from a deadly contagion, because it will disease them menally. Wrongs of the pestilent sort, such as universalism, pantheism, atheism and other importations from the pit, are spawned upon the world from this source.

"Men forget that there is a place where all the sewers of God's moral creation meet, and that place is no other than the air of your earth. The princes of the powers of evil in the air have many ways of imposing upon the earth life, and know that but for the power of God exerted in keeping his saints through faith unto eternal salvation, all in the earth life would fall of its own weight into moral chaos. But know thou that in the fulness of time this false spiritualism shall be succeeded by the good and true; and instead of being possessed and abused by harpies from the world of darkness, as is the case at the present time, people will be enlightened, cheered and saved from ruin instead of being plunged therein. The same great and blessed law that now permits the wicked spirits to approach earth life, will eventually extended to the noble and true, the religious and pure spirits from the excellent glory where God sanctifies all hearts to come to man's aid in his arduous trials; for the time must come as foretold by the prophets, when all men shall be taught of God, or by those appointed of God and there shall be none to hurt in all God's holy mountain.

"But to return to the subject of the atonement. We say man's highest conception of vicarious suffering can not reach, in the faintest degree, God's conception of that work. But know thou that there is a sense in which the pure and matchless spirit of Jesus 'tasted death for every man;' for upon him was laid the iniquity of a world, and by his stripes many are healed. That finite mind may grasp a higher conception of the atonement or the vicarious suffering principle is impossible. I announced to you the fact that the law of the higher, ministering to and making sacrifices for the lower, pervades all orders of being, from the throne of the Almighty, to the infant child, nurished upon the mother's breast. The law of vicarious suffering becomes a necessary part of the divine constitution under which all intelligences are created, because the strong only can minister to the weak—the wise only can minister to the unwise. Hence all gradations of intelligent beings are thus arranged, and it constitutes the grand order, divine. But in the gradations of beings below man this order is reversed, and the lower minister to the higher. For example, the earth with its air, its water, and its soils, ministers to vegetation; vegetation in turn ministers to the animal, and the animal in many and various ways ministers to the wants of man, and man ministers to the child. Here you will notice that as soon as the gradation reaches man the order is reversed, and the higher ministers to the lower. The toil and care of the parent ministers to the child. So the same great law reaches and ministers on its way through all the gradations of that infinite ocean of being extending from the topless throne of the Almighty creator to the inhabitants of the earth life. Thus seraphim, ministers to cherubim, cherubim to arch angels—arch angels to angels—angels to just men made perfect, and they with angels in turn minister to those who shall be heirs of glory in the earth life. Thus you discover that the law is a necessity from the very nature and constitution of all intelligent beings. Hence Jesus your Savior said, 'I care not to be ministered unto, but to *minister.*' Hence his example of the innocent suffering for the guilty is the highest possible expression of the vicarious suffering principle known to your world. He subjected himself to that condition because the necessities of a world demanded it. And though he was the brightness of the Father's glory and the express image of his person, yet he voluntarily took upon himself the form of servant, became obedient unto death, even death of the cross, that he might bring many *sons* unto *glory.*

"When that wave of suffering came from the cross, it rolled on until every order of being was brought in sympathy with man. Hence it is written, 'that angels rejoice over one sinner that repenteth.' The mission of Jesus Christ to your world was also the gift of God. The Almighty, the everlasting Father, even the Infinite Oversoul loved the world, and so loved it that he gave the dearest and brightest gem from the courts of heaven. This could not fail to interest all the subordinate orders and grades of being in the welfare of man. Thus the same law may seem to be operative in degree even in the earth life.

"When the king upon his throne feels deeply interested in the elevation of some of the poorest and meanest of his subjects, and especially if to do this the royal family makes sacrifices and self-denials, does it not the more interest all from the throne downward? Thus, perhaps, you have received a faint, glimmering conception of the operation of the laws of the atonement—the law of vicarious suffering.

"The spotless throne of the Almighty has been assailed, because he laid upon one who was mighty to save the iniquity of a world, and because he was made to taste death for every man. *Conceited folly.* They find no fault of the law operative in the family relations. Often upon the energies of a father is laid the iniquity and profligacy of an ungrateful son; and always is laid upon the mother suffering and toil that often ungrateful children may be cared for. The family is but a cluster of universes or microcosms and therefore furnishes a very proper illustration of the operation of this great law. For the father to be interested, toiling night and day, sacrificing ease and comfort for the little weak ones, stimulates all the intermediate agencies from the father, through the mother and older children down, so to feel as father and mother feels, is to feel as deeply interested as possible. But let father or mother show neglect of that little one, and it serves to brutalize the entire family.

"The cross of Christ is the central sun of the universe, and when we consider that without him there was not anything made that is made,—I say when we consider that he was before all time, that is, before this universe began, it makes the cross of Christ stand out as the representative seal of that which is the highest excellence of God's throne—disinterested love. The great distinction between good and evil consists not in the difference of intellects, but in the difference of the heart principles. The prince of evil has a mighty intellect, but not one particle of *disinterested, self-sacrificing love.* While the best representative expression of God's character is *love,* the best expression of Christ's character is vicarious suffering."

CHAPTER XXVI

Thus, kind reader, closes another very strange piece of composition from the spirit world. I say strange, because it must seem strange to every one that lofty and sublime ideas can emanate from falsehood and deception. As I said at the commencement of this chapter, the most subtle methods which the powers of the air use to induce belief in their monstrous absurdities, consist in making friendly allusions to Jesus Christ and his gospel, and in speaking very highly of some of its doctrines; perhaps give a grand dissertation upon one of them, and in the meantime weave into the frame work of the argument a species of philosophy which would undermine the consistency of the whole doctrine. Hence you will notice that the first announcment is as true as the throne of God itself. That our ideas of the atonement are far below that of the angels, and the ideas of the angels upon the subject far below the ideas which God entertains of it. But when the intelligence who dictated this article announces what he knows of atonement, he says, first, know that the human race is one mighty harp whose strings are attuned to such divine harmony, that to touch any one of them rudely, all the others vibrate accordingly. Now the super-mundane intelligence that announces such a statement to be the truth of history in regard to our race, must certainly be ignorant of the fact that one Adam broke every string of that mighty harp, and no other sound has been heard from it but a mighty weeping and wailing since the time it was so rudely unstrung.

Every man who is acquainted with human nature at all, must know that not one string of the mighty harp has vibrated in sympathy with fallen human nature, and the woes of man; but on the contrary, the only harp that could be found in all God's universe so sympathically strung as to vibrate in sympathy with the woes of man was found in heaven, and that mighty harp was Jesus Christ; and every harp belonging to our race through all time, that has in any degree vibrated in sympathy with the race, has been attuned by heaven in accordance with the swelling harmonies of the heavenly harp, Jesus.

Then, again, when the intelligence asserts that the agonies of Christ's body and soul, were shared in by every creature beneath God's infinite heavens, he certainly forgot those creatures who stood around the cross and railed on him, and also the countless millions of fallen angels and wicked men, who are doing all in their power to frustrate his designs of mercy toward the race. But then, when it is asserted that Christ's agony-thrill still goes on sweeping through the countless worlds, I think candor would say that Gabriel, God's highest angel, would hardly be expected to know how every world of countless beings was affected by the suffering o Christ upon the cross. And when this it would seem more than angelic intelligence as serts the fact of a great vicarious law, whose elements are sympathy, compensation and distribution, and that all dwellers in the spirit land are impelled to action in behalf of those on earth by the first and greatest law of the universe, *self-preservation*, or to escape the waves of pain which must necessarily reach them from the sorrowing of earth. This position is then attempted to be proved by asserting without evidence, that every sensition, painful or pleasant, experienced by any, even the most destitute, low, and degraded of the species, is shared in by all the rest of the race; and also from finding that the most sensative and finest nerved persons are always the most unhappy, therefore we infer the elements of a vicarious law.

That the finest nerved and most sensitive persons are always the most unhappy, I think is not the case; it does not accord with my experience in regard to them, for unless their surrounding circumstances are very adverse they are generally the most cheerful, hopeful and happy persons in the world. It is true we often say of them: "They have their ups and downs," but I think, in candor, their ups or ecstacies in life by far outnumber the downs or sorrowing moments of their lives.

With regard to the other position taken that every sensation, painful or pleasant, experienced by any one of the human race is shared in by all the rest, I have simply to say that it does not accord with my experience of human nature. The fact that I should suffer untold pangs in consequence of a dreadful massacre in Africa, Asia, or some cannibal island, without any knowledge of the event, is to me simply absurd. A quotation is made from the scriptures to this effect, as an additional support to this philosophical theory, viz: "Bear ye one another's burdens, and so fulfill the law of Christ." This passage I have always understood to apply to those with whom we associated, with whom we come in contact in the relation of church fellowship. Bearing burdens, or bearing with moral infirmities of others, is a Christian duty, but I think can not apply in support of the position assumed.

Another statement made to sustain the position more in detail, is to this effect: "There are seasons when men and women of the most sensitive temperaments and without the least apparent cause are plunged at once into overwhelming misery and woe, so overpowering as to be but faintly imagined by those of a different mould of organization. This comes from suffering humanity. Not a member of our race can be overwhelmed without suffering, without these sensative ones feeling it. The wave of pain goes out, and as the magnetic needle is attracted toward the pole of the earth, so these waves of pain are drawn towards these sensitive ones, even towards those whose position on the plane of the great sympathetic nerve of the universe fits to receive it. Some one else receives it in turn, but in a less intense degree, until at last a faint and tiny wave reaches the throne of God."

For the truth of this statement we appeal to facts. That a certain class of persons are often overwhelmed with seasons of melancholy, sadness, and sorrow, we know is a fact, but what causes these feelings is the question in point. I think as a general thing they can be traced, first, to a diseased body. Second, to adverse circumstances, *believed* to exist though not existing in reality. Third, to sensations of guilt arising from doing wrong.

I think all wrong feeling can be traced to one of these three sources. My experiences of the world may be summed up in this statement: All sad and sorrowing feelings invariably result from wrong doing in some way, and that on the contrary, right doing must invariably result in right feeling. This doing right includes an exalted exercise of faith in God's declaration by Paul, viz: All things work together for good to those who love Him, then their souls must rest in the quiet peace of full assurance.

We shall now notice the next statement in the order, and that it is true every mind must at once admit. As I have said these spirit communications are characterized by most exalted truths and contradictory statements and untruths. This truthful declaration is in these words: "When the suffering soul turns meekly to God, relief comes, but not an instant before. This great law that relief is realized by the soul of man, only in meek submission to his will was known in very ancient times, even as far back as Enoch, and righteous Abel. Yes, how very true, that whenever a sinful soul cries, Oh! my God, the answer comes, "child, here I am," and in every "Oh! my Father," there slumbers deep, "here my child." These sentiments are redolent of the very airs and breath of heaven. I can indorse them fully. Then follows an account of the different kinds of sensitives found in society. The statements upon this branch of the subject will strike every one as being true. After this, the author commences an extraordinary denunciation of the Atheistic Spiritualism of the present day, and promises at some future period of the world's history a true spiritualism; and instead of being possessed and abused by harpies from the world of darkness, as is the case at the present time, people will be enlightened, cheered, and saved from ruin. The same great law that now permits the wicked spirits to approach the earth life, will eventually be extended to the spirits from the excellent glory. This, I suppose, means when the earth life has progressed to that degree of sensitiveness that it will be controlled by holy and pure spirits. This is equivalent to confessing that the author of this communication is not of that number. The balance of the communication I think I could indorse, with some trifling exceptions.

The reader will bear with me if I present another specimen of spirit composition. In some respects it stands out as an unrivaled piece of composition, and though its orthodoxy may not be indorsed by all, yet the thoughts are many of them new and sublime in the highest degree. The subject treated upon, is the new birth spoken of by Jesus in John iii.

CHAPTER XXVII.

"The new birth, from one point, an angel might comprehend what it has been and now is, but not what it ultimately will be. The renewed soul enjoys untold myriads of these new births or spiritual awakenings, and only God himself can know a soul as it was, as it is, and as it will be, because God is the only being whose mind can embrace all the past, all the present, and all the future; therefore he alone can know all the depths and breadths of the so-called human soul, and especially the new birth of that soul. For man to be like his God, possessing omniscience, and to be the work, the perfect work of such a being, yea, to render him worthy the viewless Lord of infinite glory, he must be perfected. The sun shines in the heaven of spirits, just as the comparatively tiny and material one shines in and illuminates the natural universe. The spiritual does not glide into the material, but it is from and above it, just in the same sense that the meaning of a sentence is above the sound denoted by the characters which convey the meaning. You must learn to know the sublime infinitude of God's universe, understand in some measure the sublime processions of the universes of worlds, as they sweep along the eternities of time. You must receive light, life, and law from thence, to be fitted in the least degree to comprehend the measureless infinitude of the human soul, and the effect of the new birth upon that soul. Exhaustion prepares the material universe for a new infilteration of God, differing from the last, and so on forever, each procession being an eternity or cycle. Thus it will be seen by those who can grasp the tremendous thought and not let reason totter, that this amazing system of substance or matter, is after all a mere speck, a mote in the sun's rays, a mere grain on the awful shores of the stupendous spiritual ocean. Nor does all matter existing in the limitless universe bear any greater proportion to the spiritual, than an orange does in bulk to the Rocky Mountain chain. The material system moves near the center, and the spiritual waves flow out on all sides, into the grand but awful beyond. Human souls proceed from God, and return to him, that they may be judged. The new birth is the soul's first consciousness of God as a righteous, moral governor, and of God as a savior and deliverer from sin and depravity.

"Let me show you the gradations of condition from matter in its crudest form to the throne of God. First in this order is matter, rock, soil, vegetable and animal -matter. Spirit, then, may be considered the essence or soul of matter; soul the essence or main principle of spirit; monad the essence or life of the soul; think the inner principle of monad, and God the soul of think. These strange expressions stand as symbols of mighty truths. Monads are scintillations from God's brain. They are forms of thought, and are bubbles on the ever rising tide of his soul. Hence the new birth, so to speak, is the calling in of his spirit, a first step in the infinite progress of the soul toward God its maker. But the how none but himself knoweth.

"Let me try and show you again. Here is a chain. Matter is the first link, spirit atmosphere the second, soul, that which constitutes the basis of the human think principle, the third. Well, the great Oversoul flows through all these as man's spirit through his body. Man is only partly conscious; he knows absolutely nothing upon the other side of himself. He is ignorant of what life is, and of that august power which governs his involuntary self. The Oversoul flows out into the all—into the universal think—into the soul of man, while pervading all—imminent in all—is self-conscious at every point. Thus you will perceive that the subject of the new birth is most grand—is awfully sublime. Well did the Master say, 'The wind bloweth where it listeth; thou hearest the sound thereof, but canst not tell whence it cometh or whither it goeth; so is every one born of the spirit.' God is not only good, but beyond it; is not only the truth, but the all of it. He is not only power, life, and think, but beyond and above all these."

Thus closes another communication from the spirit world. Of its merits, I need say nothing. I leave the reader to exercise his own judgment, and draw his own conclusions. I will now close by answering such objections as I suppose would naturally arise in the mind of any man who has not given the subject of spirit manifestations a thorough examination. The first objection which we might naturally suppose would arise, may be stated about after this form: Is not the whole subject of spirit manifestations a delusion? Not that all the men connected with it are hypocritical and intentional deceivers, but are deluded? That is, they place too great an estimate upon certain sounds on a board or table, or they see things by the aid of a highly wrought imagination—thus being led astray?

That one or two might be thus led into the mazes of error, is conceded; but that hundreds, yea, even hundreds of thousands of all ages, capacities, and conditions, should fall into the same illogical conclusions, is absurd in the highest degree. That men should hastily form conclusions, is very common in all the every day walks of life, but that men of the shrewdest capacities, and possessed of very considerable scholarship, and even those of the very highest intellectual and moral attainments—that men of this stamp of mind and culture should *all be deceived*, is to my mind most absurd. If, then, they are not deceived, but as Paul said, " do speak forth the words of truth and soberness," it becomes us to listen to what they have to say.

That men in the flesh should be carried away by self interest is very common, but that all those who have stood up to bear testimony in order to establish the facts of spirit manifestations, should be so deeply selfish is the most unlikely thing possible. It is also true that men might be swayed by prejudice, and allow their judgments to be warped by preconceived opinions; but that so many witnesses should be thus swayed, warped or deflected from bearing correct testimony, would be a matter truly astonishing in human nature.

I therefore must decide that the man who could come to such a conclusion with the premises above stated must be a fit subject for the mad house, or the lunatic asylum. That there are such men is very certain, and that they will, without a moment's reflection cry out "humbug," is also certain; and that they will also remain outside of an assylum is as equally certain, but that they have any weight in directing or forming public opinion, no one will for a moment admit. They are warts upon the body politic; they are a fungus growth, and must, like all mushroom productions, perish before the high noon of public decision.

It may be proper to repeat here another very trite objection—I allude to that which is so common among ministers of the gospel, to-wit: That spirits of the wicked and righteous dead are either in hell or heaven, and therefore cannot be here about our earth.

To this we answer, that spirits of both characters are in the bible represented as not only being here about the earth, but also of influencing men in the flesh. Notice that revelation commences by the mention of two classes of spirits, the one called cherubim, the sons of God, the morning stars, as found in this expression of Job: "Where wast thou when I laid the foundations of the earth; when the morning stars sang together, and all the sons of God shouted for joy?" The other class is included under the head of old serpent, which is the devil. Now, if at the beginning where revelation opens, the two classes are mentioned, and this distinction is kept up all the way down through the writings of the prophets, inspired poets, apostles, evangelists and teachers, then it must be certain that the two classes of spirits exist about this earth.

When it is said that angels are the spirits alluded to, we answer that the spirits of just men are spoken of as being interested in matters upon the earth; that among these spirits of just men are mentioned Moses and Elijah, who appeared and talked with Jesus upon the Mount, and many of the saints who arose at the resurrection of Christ, and appeared unto many, besides John the revelator mentions the fact that one of the old prophets appeared and spoke to him, and in the excitement of the moment was about to fall down and worship him, but he quickly said, "See thou do it not; I am thy fellow-servant," undoubtedly conveying the idea that those old saints serve in the cause of Christ as well after death as before that event took place.

And when it is said that devils are the only class of spirits alluded to in the Bible, under the head of wicked spirits, we answer, if that is the fact why should they be classified and spoken of as being different, by bearing different appellations, the expressions "familiar spirits," "lying spirits," "seducing spirits," "unclean spirits," and the like? If all these expressions were one and the same class of spirits, it is a very strong use of language. But in the expression seducing spirits, and doctrines of devils, the two classes are contrasted—it can not be that only one class of beings is intended.

But the objector will say: It is not possible that the human mind can become so sensitive as to be able to receive an idea from a spirit without the use of the vocal organs.

Spirits speak to one another by thinking the thought distinctly out, just as a man would do if thinking to himself. There is an inconceivably fine element in the air which carries the thoughts of spirits just as the air carries our thoughts—called the thought atmosphere. All men could speak to one another without the use of the vocal organs, provided their minds were sufficiently sensitive. How often it is the case that sensitive persons in earnest conversation perceive the thought before it is expressed by their opponent. How common is the remark "I knew you were going to say that"—alluding to the fact that they had already caught the idea.

Females are more sensitive than men. You will often find women so sensitive that they experience pleasant or unpleasant sensations whenever they are thrown into the society of the good and pure, or into that of the wicked or impure. A lady upon one occasion declared to me that she could in a moment realize a difference of character in different persons whenever they by chance came into her presence. It is no uncommon thing for people to say "I formed a dislike to such and such a one." Robbers and persons of the baser sort have often been detected from this very source.

There is an inconcievably fine emanation passing off, so to speak, from the soul and body of every person, and this emination (call it magnetism, if you choose), partakes of the nature in a degree of the moral character of the person it proceeds from, and whether you realize it or not depends upon your degree of sensitiveness. How common is it for hunters to take advantage of poor animals, because they must leave a small portion of their emanations in their tracks? Dogs, or many other animals, are sensitive to perceive these infintissimal particles, and can, by means of it, pursue at great speed. Think how little of this emanating substance must be left in the track of a man when his dog can easily follow it an hour or more after he has passed!

From these facts we may learn that when it is asserted that a spirit can speak to a man in the flesh, or in other words can project his thoughts into his mind, it depends only upon the sensitiveness of that mind. Some minds, like the weeping willow, are sensitive to the slightest impressions, whilst others, like the branches of the sturdy oak, are shaken only by the mighty winds. That men have heard spirits speak to them is the positive declaration of scripture in very many instances. Elijah heard it as a still small voice. Paul heard distinctly the voice of the spirit of Jesus, but the men who were with him heard not the voice which spoke to him. From this we may learn that no vocal sounds were made when Jesus spake, else the men who accompanied Paul would have heard it. It must therefore have been addressed to Paul's mind as a still, small voice.

How often do we hear Christians, in relating their soul experiences of God's pardoning mercy say, "A voice as audible as if spoken to me by a person in the form spake to me, saying thy sins which were many are all forgiven thee," or "My peace I give unto thee," and other similar expressions? Is this all imagination? Surely it can not be. In my own case, I remember that at the time my soul was seeking very earnestly to realize a consciousness of God's forgiveness, that a voice came to my inner being saying in these emphatic words, "The way of salvation is repentance toward God, and faith in the Lord Jesus Christ." The effect of these words so impressibly spoken was to make me exclaim aloud, "There it is—the whole of it in a nutshell." A friend occupying the room with me asked, "What is it?" to whom I made this remark: "I wonder I never understood the meaning of that scripture alluded to above, before."

Instances might be multiplied innumerably confirming the fact of persons hearing not voices, but words spoken to the mind, which made as deep an impression as if spoken by a human voice, audibly.

"But," says a questioner, "Can you not give us briefly, a few of the evidences which have determined your mind to conclude that spirit intelligences speak to you?" I would say, first, that I have all the evidence that any man could have of the existence of a friend or stranger, who might stand outside of his house in a dark night. I think these evidences would be conclusive if, the stranger or friend was able to answer every question you might propound to him promptly, upon a great variety of subjects, and if instead of simply answering your questions, the person in the dark should begin to ask *you* questions, questions, too, of a very startling nature— questions propounded with shrewdness, with wisdom, or in jest, or mirth. But if in addition to giving intelligent answers to your every question, and also propounding intelligent questions to yourself, the person outside in the dark should say to you, "You are a liar and I can prove it," and then should go on and give the particular proofs, or should say that "you were a "political scalawag," and should give the evidences of that also, and so on, charging you with many delinquencies and giving the proof when demanded, I say when this should be done, you would begin to conclude, at least, that some body was outside of your house, and a somebody who was pretty well acquainted with your doings.

But supposing, being still in doubt, you should resort to another method—and you should begin to apply epithets to them, and they should in turn berate you soundly, using language most insulting. Or, supposing you commence in a strain of flattery, and they in turn should apply the soft sodder to you with such unmerciful thickness that you are glad to beat a retreat. But in case a doubt still lingered in your mind of the identity of the person outside, we might suppose you to resort to this last test—I will raise the window, and if he strikes me when I provoke him, I will know then of a certainty of his identity.

I will say then, in conclusion, that all the tests mentioned have been resorted to, to determine the spirit intelligences about me, and particularly the ones mentioned last. I have oftened applyed abusive epithets to them as a kind of a last test of their identity, and found at times, they would receive it as a joke, using other epithets in return; but when persisted in, they would turn upon me sharply, saying, "Dry up that kind of talk or we will teach you a lession which you will not be slow in forgeting," and then would commence a series of perhaps, a half dozen, or a half score even, of electric shocks as severe as that given by an ordinary electric battery. This test I have tried so often and with such unpleasant results, that I am disposed to treat these spirit beings precisely as I do those in the body, with kindness and respect. Occasionally they shock me without provoeation, and when asked the reason for so doing, receive as a reply—"We only want you to know that we are about."

But in saying that you believe in spirit manifestations do you not lay yourself liable to be called a spiritualist? I think not. Spiritualists, so-called, in addition to their belief in the physical and other demonstrations of spirit intelligences, embody a system of philosophy which goes in the world under 'the name of Pantheism. This system discards the personality of God, and all divine revelation from him, and announces that all men have the germ of an endless, glorious life within them, only to be fully realized it must be cultivated.

Every reader of this volume must certainly know that I deny all this pantheistic philosophy, knowing full well the source from whence it emanates. To say that a man is a spiritualist because he believes in the fact of spirit manifestations, would be as unwise as to charge Jesus with being a pharisee, because with them he believed in the existence of angels and spirits.

I would say that I have great charity for spiritualists. I have reasons for thinking that in general they are sincere, and desire to be guided by facts and evidence, but placing too much confidence in what the spirits say to them they are led off into the wildest theories and absurdities. I hope I am not of those who are so unwise as to say they will not believe amidst abundant proofs, or of those who believe without any, or without sufficient proofs.

I think it is the highest act of wisdom to

disabuse our minds of all prejudice, to collect facts carefully, to discriminate closely, and then to decide candidly. That I may have some prejudices, and even biases of mind, is altogether probable, but that I have tried to disabuse my mind of them I think the reader will do me the justice to believe true.

If, upon farther investigation, I shall obtain a sufficient number of facts to warrant me in a change of my present views, rest assured they will be changed, but not until then. But I would say to every lover of truth that they are bound by every principle of right and justice toward a fellow man to aid in the promotion of correct opinions. If, therefore, they will address me by letter, stating their views and communicating any facts, I shall be happy to receive their addresses and answer, or if they desire to propound any questions they shall be carefully considered, and an answer returned.

APPENDIX.

It may have occurred to some of the readers of this volume that corroborative facts would be of much force in helping the mind to arrive at just conclusions in the premises.

I would say, in reply to this probable suggestion of some readers, that I am in possession of many facts connected with the experience of several persons which bear with much weight upon the important question before us. My experience being of a character altogether strange and bordering on the marvelous, I have therefore been led to make especial inquiry into the experience of others in order to asertain the fact of corroborative evidence.

The first case brought to my notice is that of a Mr. K., at present a resident of this State, but formerly a resident of one of the more eastern States. He stated to me that he was formerly a member of the Methodist Church; had experienced what is termed a passing from death unto life, consequently enjoyed religion. He stated farther that his mind became interested in the strange manifestations, which prevailed to a considerable extent in the neighborhood where he formerly lived, known as spirit manifestations. That in process of time he became a trance speaker; that is, as he states, he felt a strange power come over him, which at times controlled his vocal organism, and he was made to utter even in the meeting house strange sentences and give expression to still stranger thoughts while in the act of praying. This at first passed off as eccentricity, but occurring again and again, it was termed derangement. Finally, to cap the climax of the marvelous and the strange, he, upon a certain occasion, arose with all the dignity of an orator and delivered an address upon the North Star! This, with his previous eccentricities, confirmed many in the opinion that he was insane.

At length his mind became so impressible that the spirits, he says, could speak to him even as a friend speaks with a friend. Reposing some confidence in these, as he thought, spirit friends, he was led to follow their advice, but by so doing he was made to *do* as well as to say at times many strange things. Finally, by frequent deception practiced upon him, he lost all confidence in the integrity of the spirits, and was led to abandon their advice as of no more weight than the advice given by a set of wicked, designing men. Though years have elapsed since the above circumstances occurred, yet the spirits exert over him still a very surprising power. At times an Indian spirit of great power desires to control him, and though he uses many exertions to keep himself positive to his influences, yet at times he gets partial control of his vocal organs and as a result strange sounds are produced.

But what seems the most surprising—although he disclaims utterly the reposing of any confidence in what the spirits might reveal to him as spirits—he yet receives and subscribes to the so-called spiritual philosophy, which he must know all came from spirits. He expresses the opinion that spiritualism in its moral aspects is very weak; that it needs to be baptized with the Holy Ghost power to become a potent agency for good in the world.

Another case which has come under my notice, is that of a lady, a Mrs. C., residing at this time in the State of Indiana. She became a subject of abuse by the "so-called spirit friends" by first yielding to their influence in the following manner: She was sitting at the table upon one occasion in the act of writing, when she discovered her hand to move involuntarily, which occasioned her much surprise. After repeated tests she became conviced that some foreign influence must have something to do with it. What convinced her more than any other one thing that the power to move the hand without the volition of the mind originated from spirits was the fact that the hand did not move around without any regard to form, but was guided to make the most elaborate pictures, so that she became an adept at fine pencil drawing. This seemed to her the more remarkable as she never had, previous to this time, realized any inclination or ability to sketch fancy pictures. This practice of sketching pictures with a pencil was continued for a time with very considerable interest, but at length she became wearied with the practice and made the attempt to abandon it altogether. This she found to be more difficult than she at first expected, for her spirit friend, as she called herself, felt exceedingly anxious to continue her course of mediumistic drawing. At a certain time each day there was heard certain plain, distinct wraps, indicating that her spirit friend was ready to commence operations, and whether the subject could devote the time or not still the wraps were persisted in until the subject yielded. This persistent course, as might be supposed, became exceedingly annoying to the subject, and after a time matters between them came to an open rupture. The spirit became vexed and the subject also became vexed and persisted in her refusal to sit for purposes of drawing. Thus matters continued for many days —neither of the two determined minds could be induced to yield. One evening as the subject was sitting at her usual avocations, surrounded by some of the members of her own family and one or two of her neighbors, she was forcibly seized and controlled to pick up a small missile and throw it violently toward some one in the room— looking in the mean time like a very demon. This unexpected move alarmed those present, who ran violently out of the house and took refuge in the house of a neighbor, the subject following at the top of her speed being as much frightened as those whom she pursued.

This, the lady informed me, was her first experience in mediumistic control, and she hoped it might be the last of *that* kind of control. I saw her afterward upon several occasions pass into what is called the trance state, and speak in a most eloquent and impressive manner. These conditions continued for some time, when she became like myself, so impressible that she could distinctly hear the spirits as they spoke to herself and others.

But these spirit-friends, as she called them, like many in this life, imposed upon themselves restraints until they had measurably won the confidence of their subject. Then gradually they began practicing upon her the gravest deceptions, until finally she lost all confidence in them. She has hopes, as she remarked to me, that good spirits may get control, and then matters would proceed without disturbing influences, but I have never learned that the good spirits came; on the contrary, upon an occasion when a mediumistic friend was making her a visit, the friend was controlled to sing what seemed to be an Indian song, and this lady was controlled to dance for nearly an hour.

Thus I might go on multiplying cases that have come within the range of my own observation. I will barely allude to two other cases, and pass to other considerations. These cases all bear upon the point which I wish to establish, that whenever spirits have gained control of a subject they have invariably deceived and otherwise abused them.

A Mr. D. and wife became what is called personating mediums. The spirits took entire control of their persons, and caused them to imitate the peculiarities of character as seen manifested among the various nationalities. After allowing the spirits to use their persons for the purpose alluded to for a time, they found to their sorrow that they had placed themselves under the power of "strange friends." The spirits began by taking liberties with their persons, and finally ended in their not being able to sleep in a bed, for their strange spirit-friends could, at their pleasure, and did for a long time, pull them both out of bed. This became such an aggravation that they finally, to escape the annoyance, made their bed upon the floor.

and continued to do so for more than six months. Finally, by rendering themselves positive to the influence of their too familiar spirits, and by not allowing them to gain in any respect spirit control of their persons, they were relieved in a great measure from the disturbing influences of their tormentors.

A Mr. R. also became a personating medium, and such entire control did they gain of his person, that they caused him to run up a flight of stairs with almost the speed of lightning, and then, for perhaps an hour, caused him to make every possible gesticulation, personating a great variety of characters. His brother remarked to me that he felt no alarm for his brother's safety, though the spirits seemed to handle him very roughly. What injuries to the person of this young man may result from the spirit's control has not as yet transpired; but my opinion is that they will be similar to those experienced by the others I have alluded to.

Since I have begun to write out my strange experience of spirit control, I have been lead to read the experiences of Emanuel Swedenborg, and I am convinced that all his visions of heaven and hell were psychological images, photographed into his mind by the spirit's operation. Swedenborg gives it as his opinion that he was disembodied—that is, his spirit was permitted to leave his body and go out into the spirit world and see and talk with persons who had deceased. In regard to this I am very positive that he was mistaken. His mind pictures correspond exactly with my own, and when he thought he was out of the body he was only partially or wholly psychologised by a spirit who willed him to see a thing and he saw it. I have been in the same condition hundreds of times, and I know that my spirit never was put out of or detached from my body.

I will give one or two examples taken from his Christian religion. On page 524, paragraph 794, he expresses himself thus:

"From the things seen for so many years I can relate the following: There are lands in the spirit world as well as in the natural world, and that there are also plains and valleys, mountains and hills, and likewise fountains and rivers. That there are gardens, groves, woods, cities, palaces, houses, gold, silver, and precious stones; in a word, that there are all things whatsoever that are in the natural world; but those in heaven are immensely more perfect." (The reader will notice here.) But he says the difference is that all things that are seen in the spiritual world are created in a moment by the Lord; but that all things in the natural world exist and grow from seed."

I ask the candid reader if it would not be more reasonable to believe that what he saw as lands, trees, houses, &c., in the spirit world, were mind pictures rather than real, veritable things, especially when he says they were created in a moment by the *Lord*, who might be no one else but the spirit who controlled him, and who palmed himself off as the Lord, just as in my own case.

On page 202, paragraph 277, he says: "One day, in the spirit, I rambled through various places in the spirit world in order that I might observe the representation of heavenly things which in many places are there exhibited. In a certain house where there were angels, I saw great purses in which silver was stored, and because they were open it was perceived as if every one might take out, yea steal the silver their laid up. But near these purses sat two youths who were guards. The place where these were deposited appeared like a manger in a stable. In the next room were some modest virgins, with a chaste wife, and near that room stood two children. After these things were seen I was instructed that by them was represented the natural sense of the word in which there is a spiritual sense. The great purses full of silver signified the knowledge of truth in great abundance."

In this relation who can not see that Swedenborg's spiritual world was all in his own mind, a mere mental picture, just as the Congress of the United States was seen by myself in full session? I think every unbiased mind will agree with me that they are identical.

On page 236, paragraph 325, the reader will find these words: "Once at the dawn of day, when I awaked from sleep I saw before my eyes, as it were, spirits in various shapes; afterwards when it was fairly morning I saw fatuous lights in divers forms, some like sheets of paper full of writing, which being folded together again and again at length appeared like falling stars which in their desent vanished in the air, some like open books, some of which shone like little moons, some burnt like candles, &c. From these appearances I conjectured that below these meteors stood those who were disputing about imaginary things which they esteemed of great moment. For in the spiritual world such phenomena appear in the atmospheres from the reasonings of those standing below. Presently the sight of my spirit was opened to me and I observed a number of spirits whose heads were encircled with leaves of laurel, and who were clothed in gowns adorned with flowers, which signified that they were spirits who in the natural world had been renowned for their fame and erudition, and because I was in the spirit I drew near and mingled myself in their company."

The reader will notice that this vision was in the morning, even after full daylight. He says he saw spectres in various shapes, and fatuos lights, &c. Similar appearances, and also very different ones, I have seen nearly every morning for years, even in broad daylight. I have asked others if they saw them, and when told that they did not I knew it was a mental picture. Sometimes my room would be covered with heavy pannel work, at other times with medallion work, then, again, some other mornings the room would be highly ornamented with the most elaborate stucco work brought out in bold relief, then, again, other mornings it would be changed to fresco, and raised laquer work, the latter was of a tan color, the fresco was variegated with the appearance of many colors.

I satisfied myself that all this resulted from being paritaly mesmerized by spirits, and then by will-power on their part photographed into my mind so as to appear like actual, real objects. I could dispel the whole vision by manipulating my face, and making a sudden effort of the will.

I here make an extract from the "Banner of Light," a Boston spiritualist paper, dated January 27, 1866. The question is asked a spirit, "How do you materialize yourselves, or render yourselves visible?" The answer is in these words: "You see we have got to hitch on and draw out the machinery to materialize ourselves. It is done the same way that some persons in India tell you they can make a tree grow right up before your eyes. They do it by the same thing; that is, by throwing their whole magnetism upon a certain spot, and through the very *elements of will up grows that tree before your eyes to any hight you wish.* That's materializing what's in the ground, it's the condensing process."

Notice, the spirit says here, "through the very *elements of will up grows the tree.*" But it's only in the minds of the person seeing the tree that it goes up—nothing more.

A writer in the "World's Crisis," uses this language: "Let it be borne in mind that all the spirits and spiritualists, so far as we know, are unanimous in teaching that the spirits operate just as a mesmerizer does. Having found that a mesmerizer acts by the power of *his will*, we are prepared to understand the wonders of spiritualism.' He quotes from the Banner of Light, of April 9, 1864. The question to the spirit is as follows:

QUESTION.—"Is the flower as tangible to the disembodied in spirit life as it is to us?"

ANSWER.—"Flowers in the spirit land take the form of beautiful thoughts."

QUESTION.—" Are there not real spirit flowers, such as we are accustomed to seeing here."

ANSWER.—"No, there are not; Pardon us if we have

swept away the pleasant illusion; but some one must do this sooner or later."

I think every candid, unprejudiced mind who reads these quotations, will come to the same conclusions which my own mind has come to: that the spirit world, so beautifully described by some spirits, is a mere fiction—a heavy draw upon the imagination—especially that spirit world which is within the limits of the earth's atmosphere. I think the spirit who gives the answer to the question, "What end do the wicked and unsaved of earth life subserve?" in giving her first experience in the spirit land, comes the nearest to the facts in the case. These are the words used: When first I emerged from the body into this strange abode, the receptacle of the moral filth of earth, I was was first overwhelmed with grief, then with rage, then I sank down in despair; I cursed my existence, the day of my birth, and then my *maker*. I looked up and around me, the same heavens were indeed above my head, the same earth beneath my feet, but all else how changed. Instead of the heavenly glory into which I flattered myself I should emerge, I was surrounded by the same brick walls, the same old shabby out-houses, the same front yard, trees, garden, and orchard. Instead of being surrounded by angel bands whom I had expected as an escort to heaven, I was surrounded by beings horrid looking in human form."

www.ingramcontent.com/pod-product-compliance
Lightning Source LLC
Chambersburg PA
CBHW020300240426
43673CB00039B/653